Connected Digital Devices in Health

Computing and Connected Society Set

coordinated by
Dominique Carré and
Geneviève Vidal

Volume 4

Connected Digital Devices in Health

Edited by

Dominique Carré
Geneviève Vidal

ISTE

WILEY

First published 2026 in Great Britain and the United States by ISTE Ltd and John Wiley & Sons, Inc.

ISTE Ltd
27-37 St George's Road
London SW19 4EU
UK

www.iste.co.uk

John Wiley & Sons, Inc.
111 River Street
Hoboken, NJ 07030
USA

www.wiley.com

Library of Congress Control Number: 2026931577

British Library Cataloguing-in-Publication Data
A CIP record for this book is available from the British Library
ISBN 978-1-78630-965-5

The manufacturer's authorized representative according to the EU General Product Safety Regulation is Wiley-VCH GmbH, Boschstr. 12, 69469 Weinheim, Germany, e-mail: Product_Safety@wiley.com.

Contents

Presentation of the Authors

Yanita Andonova

A university professor and director of the LabSIC laboratory, her research work relates to communication within organizations, acknowledgment of the individual at work, requirements for creativity, creative industries and the challenges associated with digital technology.

Dominique Carré

An emeritus professor, his research relates to digital innovations and communicational devices in a context of intensive, rapid and hyperconnectivized industrialization. He is co-editor of the online journal *tic&société*.

Étienne Hien

A PhD candidate in Information and Communication Sciences at Sorbonne Paris Nord University, within the RHU PsyCARE/GHU Paris research project, his PhD explores how young people and professionals appropriate digital platforms to address mental-health issues.

Christian Papilloud

A university professor, his teaching and research are in the fields of social theory, especially relational sociology, sociology of culture and integrated mixed research methods based on computational social science.

Sarah Sandré

A lawyer at the health data warehouse facility at the AP-HP public hospital (*Assistance Publique – Hopitaux de Paris*) and historian of sciences and contemporary techniques (PhD), her research is in the area of digital health. She leads a working group on this subject, supported by MSH Paris Nord.

Geneviève Vidal

A university professor, her research concerns digital uses in a sociopolitical dynamic. She is the president of the Creis-Terminal association.

Introduction

To summarize how connected objects and digital devices are being deployed in healthcare, and to understand the underlying strategies which have structured and organized the field of health and healthcare since the end of the 20th century, it would have been unthinkable and inconceivable, with presentism and generalized technologism, to analyze it without contextualizing it using a sociohistorical approach. Therefore, instead of developing an approach based exclusively on this innovation, we felt it necessary to consider a broader dynamic and an extended timeframe, in the healthcare sector and in the field of health in broader terms, while adopting a communicational perspective. With an emphasis on exploring different perspectives and a multi-disciplinary approach, the publication addresses a theme which constantly reappears in discursive productions: modernization in health as an indicator of the promise of living life in better health.

To assess the integration of this type of technology into society, Dominique Carré (Chapter 1) proposes to start this publication using a critical approach, to draw up a framework and, in a certain way, a reading grid which should encourage better understanding of the way connected objects from "well-being" to "health" have been rolled out in

Introduction written by Dominique CARRÉ and Geneviève VIDAL.

recent years, extending or reinforcing the idea of "moving closer together through detachment" within care practices and the rapid development of the so-called doctor–patient bond, a communication configuration connecting the doctor to their patient. The arrival of new actors in the healthcare field, most often the communications industry, start-ups which until then did not officiate in the health sector and which are now occupying an increasingly significant position, highlights the growing importance of using a multi-directional network embracing the data economy.

Doctor–patient relations are at the heart of the work carried out by Christian Papilloud, Geneviève Vidal and Yanita Andonova (Chapter 2). They explore the challenges related to automation and connected objects in medicine based on an analysis of the practices of doctors treating patients suffering from sleep apnea. The patients use a medical device which records data during their sleep, which is then accessible to doctors and patients via applications. The analysis models the statements collected from 20 doctors to understand the uses of these connected devices and their consequences on patient relationships. It underscores the increasing importance of technology in medical practice and highlights the impact of these connected objects on the relational dynamics between doctors and patients. It indicates that the data produced by patients provides doctors with the resources for a population-based approach to patients and their illness, ideally allowing the establishment of diagnoses and treatment plans to become automated.

It is helpful to examine the case of mental health to understand the extent of the information technology invasion, leading to remote patient care. The issue of confidentiality regarding sensitive data is a concern for healthcare providers in their interactions with patients. Nevertheless, the health sector believes in the contributions of digital technology to continue care, provide information and facilitate follow-up assignments. The contribution of Étienne Hien (Chapter 3) is eloquent on this subject, in particular concerning the "impact" of digital technology in psychiatry. Patient monitoring is essential, and alert or virtual game devices, as well as data collection for treatments, are proposed, in addition to follow-ups through digital social media. The author, who advocates for complementarity between digital technology and doctor–patient human relations as opposed to substitution, begins a debate about technological solutionism. At the same time, inequalities and risks associated with the use of sensitive data emerge, despite regulations governing their protection.

Sarah Sandré (Chapter 4) questions the multiplication of experimentation with connected devices in healthcare. In her chapter, she discusses how the government, in the same way as public authorities in a broader sense, has rolled out an incentive-based policy and many related mechanisms: funding, calls for project proposals, expressions of interest, the creation of third-party sites (incubators, acceleration programs), partnerships and working with doctors or hospital departments. However, Sandré observes a shift between "technoscientific promises", which connected objects benefit from, the actual efficiency of specific devices, and the difficulty in their implementation and deployment within hospital departments (legal risk management, appropriation of the technique, information and training of the patient, medical protocol, organizational procedures, data protection, etc.). The author also notes that when it comes to changes related to technical innovations, humanities, social sciences (sociology, information and communication sciences, etc.) and legal sciences are curiously absent, whereas they would undeniably provide real added value.

The case of nanomedicine offers valuable insights for extending thinking on the healthcare sector during its technologization, according to the sociological macro-relational approach applied by Christian Papilloud (Chapter 5). Nanomedicine (for new medications and new treatments), based on mass data relating to pathologies, in the context of medical research and the detection of pathologies to perform economic health policies, is another promise of improvement to individuals' lives. The critical approach to this situation reveals a concept: satellization arising from the collaboration between science and the economy, founded on technological innovation and supported by strategies of allied players. The European Technology Platform is an illustration of a partnership between science and the economy in the health industry. Through the analysis of the Translation Advisory Board (TAB), we can grasp the challenges of the transition from nanomedicine to precision medicine and to Key Enabling Technologies (KET), a satellization of the economy carried forward by the political sphere, with the support of digital media (data suppliers). These strategies carry forward the health system, assessing personalized treatments based on models of pathologies carried on by digital technology.

Personalized and predictive medicine, made possible thanks to objects connected to computer networks and artificial intelligence, contributes to the widespread collection of data in the healthcare sector of a society permeated by information technology. Accelerating, intensifying and complexifying the computerization of society appears to justify the adoption of a strategy aiming towards digital, as well as political, transformation of the health sector. The belief in technologies using connected devices undermines the trust in solutions for public health and care. Yet, these technologies are both intrusive and ambivalent, in the sense that they facilitate care and medical monitoring. However, they generate at the same time sensitive data exposed to reticular flows, within a context of a digital data economy. Certainly, precautionary and ethical behaviors are the recommended positions to take, but the risks are immense. Some consider the benefits and the disadvantages, but sociopolitical challenges exceed this accounting-based logic of advantages and disadvantages. Indeed, the use of connected technologies in healthcare and doctor–patient interactions is subject to conditions, according to Geneviève Vidal (Chapter 6), and the datafication of the health sector leads doctors and patients, who are not in control of the situation, to be caught up in a strategic device.

1

A Sociohistorical Approach: Moving Closer Together Through Detachment from Care Practices

To analyze the conditions surrounding the introduction and spread of connected objects in healthcare (i.e. connected digital objects), we considered it necessary to avoid developing an approach focused solely on this innovation. Instead, we chose to recontextualize it by situating connected objects within a broader dynamic and a longer timeframe, not only within the healthcare sector, but more broadly within the field of health. To this end, we adopt a communicational perspective that accounts for a recurring theme in discourse: the modernization of healthcare as a promise of better health outcomes, whether through production, prevention, monitoring or support for care.

To describe the conditions for integrating this type of technology, our first step is to adopt a broader lens and revisit the 1980s to understand how what we term "moving

---

Chapter written by Dominique CARRÉ.

closer together through detachment[1]" emerged and evolved.

Next, we examine how, during the 2010–2020 decade, connected objects – in a broader sense – became part of, extended, challenged or failed to align with this process.

Finally, we explore how connected objects reinforce this process of moving closer together through detachment – notably by introducing more sociotechnical mediations – limiting the production of sensitive data while aligning with a usage-based logic that gradually took shape during the development of remote health services.

What should be understood by "moving closer together" through "detachment"? It is the promise of remaining as close as possible to the patient while care practices and health organizations simultaneously withdraw, establishing a sociosanitary distance from bodies and individuals.

What is proposed here is, in a sense, a sociological interpretation that combines "moving closer together" and "distancing" in healthcare – highlighting the role of digital communication. It should be noted that this concept emerged from research conducted with our colleague Robert Panico and was later explored in a French follow-up publication titled "Informationalization in Healthcare or Moving Together through Detachment? Elements of an Analysis of the Economic and Organizational Management of Healthcare Distance" (Carré and Panico 2003). That study sought to describe "the terminological indeterminacy that arises from the combination of two terms that appear both contradictory and simultaneous: beyond the spatial metaphor, it is a

1 This concept refers to the promise of being as close as possible to the patient (in rhetoric and policy) while healthcare practices and organizations are moving further away (closure of local units, such as maternity wards, or consolidation of hospitals), thus creating a significant socio-medical distance from bodies and individuals.

matter of interrogating this semantic ambiguity…" (Carré and Panico 2003, p. 102).

There were two objectives to this research. On the one hand, we deliberately posed a somewhat provocative question about detachment, taking a contrarian stance towards a reform agenda which, under successive governments, aimed to link system modernization with improved access to care and closer proximity to healthcare through the territorial reorganization of health services and the new coordination of healthcare providers centered around the patient. On the other hand, we sought to examine a less frequently discussed aspect of modernization – less celebrated because it is more pragmatic – namely, cost control, which is now less of a taboo, as the rationalization of actions and increased oversight of care practices have become standardized.

1.1. The shift to ambulatory care: digitalized as a structuring framework

Faced with rising public healthcare expenditures, which in France account for nearly 10% of GDP[2], public authorities, among others, came to view the sector as in need of reorganization to restore order, as it was becoming ungovernable (Julien et al. 1987). At that time, public authorities began to adopt a stricter, accounting-based logic, leading to a more quantitative and enduring vision (Morel 1997), which has only grown stronger. This drive for reorganization centered around what became known as the "Bostonian" model, and the following series of objectives: hospital closures or consolidations, fewer doctors, shorter hospital stays, increased patient involvement and expanded outpatient surgery, among others.

2 Whereas they only represented 7.6% of GDP in 1980 and 4.4% in 1960.

All of this was captured in the term "shift to ambulatory care," which rested on four key pillars (Carré and Lacroix 1999): the establishment of technical networks in the healthcare sector; the creation of a computerized patient record, initially shared, then more recently integrated into an online personal health account; more collaborative care pathways involving healthcare professionals; and finally, increased patient autonomy and task externalization. These developments encouraged the entry of new actors into the healthcare field, allowing for short-term experimentation and the gradual rollout of remote services (e.g. remote monitoring, remote consultations, tele-diagnosis, etc.).

Here, the concept of remote services should be understood in a broad and generic sense, often situated at the intersection of market-based and non-market-based models (Carré 2001). These changes are thus part of a dual process: the implementation of healthcare administration and oversight[3] through the establishment of sociotechnical mediations between healthcare providers and patients.

The sociosanitary crisis, as Carré and Lacroix (2001a) remind us, is politically driven. The arguments put forward are not based on a need to respond to malfunctions in the healthcare sector, but rather stem from the view that health costs (i.e. budgetary concerns) are too high and, as some

3 By way of examples: Various regulations have been enacted to reform the French healthcare sector. These include: the law of July 31, 1991, which obliges health establishments to analyze and evaluate their activities – a law that followed the adoption of the Medicalization of Information Systems program (PMSI) in the late 1980s; the law of January 4, 1993, which established coding for medical procedures and pathologies, thereby implementing a generalized, standardized assessment of care practices aimed at controlling healthcare spending; a Social Security reform that requires doctors to digitize their practices to facilitate data exchanges between patients and health insurance providers; the circular of May 10, 1995, which extended the PMSI nationwide; and the ordinance of April 24, 1996, which reformed both public and private hospital care to promote greater cooperation among institutions.

claim (specialists, experts, public authorities), threaten to slow France's economic growth and competitiveness. As a result, budgets must be cut and streamlined[4]. The solution is imposed without public debate, and implemented through the technical networks of healthcare actors.

Four communicational strategies will be implemented, more or less chronologically:

– de-singularizing the healthcare sector to make it more ordinary, with the aim of treating it like any other sector – allowing comparisons that highlight its lower productivity;

– assigning individual responsibility to healthcare actors, using approaches that induce psychological discomfort to promote personal changes in attitude or behavior. In other words, individuals are held accountable, particularly in cases of medical nomadism or excessive medication use;

– imposing regulations, sanctions or threats to incentivize those who are reluctant to comply;

– promoting telemedicine. This final point highlights technical achievement and noble goals – such as improving healthcare quality, humanizing medicine and enhancing patient autonomy – while downplaying less acceptable, more industrial or labor-intensive goals that are harder to justify to the public and healthcare professionals.

1.2. Innovation and technology: the position given to digital ICT

The healthcare sector, and particularly hospitals, is a major driver of innovation in our societies, adopting emerging technologies across many disciplines (surgery, radiology, hematology, among others). It is worth noting that hospitals were among the first institutions to computerize

4 This process is taking place in most Western countries because healthcare expenditure is rising significantly faster than GDP.

their administrative systems and exchange data electronically. These technical innovations extend traditional methods of examining the human body while also improving and streamlining them to enable more accurate diagnoses, as seen in the evolution of radiography, particularly ultrasound, MRI and CT scans.

These tools have gradually become diagnostic aids (expert systems of varying complexity), but a key shift occurred at the turn of the 21st century with the rise of digital information and communication technologies (ICTs). Their introduction began to influence not only care delivery but also administrative and financial systems (healthcare economics), and even served as communication tools – through secure messaging platforms, digital portals and other interfaces – between institutions, professionals and patients.

All of this has had a significant impact on operating methods, team dynamics and modes of cooperation, progressively integrating into patient interactions. The phenomenon intensified throughout the 2010s and particularly during the Covid-19 pandemic.

Thus, digital information and communication technologies hold an important position in health modernization, since they are both the result of reform; modernization requires digitization and the use of networking by stakeholders and, in turn, they drive reform by introducing new actors and strategic logics that were, until recently, foreign to the healthcare sector.

According to public authorities at the time, the goal of these techniques, among others, was to reshape the delivery of care around the patient, ensuring the closest possible proximity (Carré and Panico 2003). This process of social digitization even extends beyond traditional care settings (clinics, hospitals) to reach medical practices where community-based medicine is provided.

The modernization deemed necessary thus encourages networked healthcare, enabling information exchange and the expansion of so-called remote services, contributing to a new vision of care access and practices, in which health becomes integrated into the so-called digital society.

All of this is accompanied by the slogan "Make information travel instead of patients," which promotes the de-territorialization of care. While the healthcare sector faces this transformation, it is not alone; automation, rationalization and the externalization of informational and communication networks are part of a broader, cross-cutting social logic that has gradually imposed itself on other sectors, including businesses, government agencies and local/regional authorities.

1.3. Distance and proximity: medical biology and medical imaging

While technical–organizational tools such as remote monitoring have impacted doctor–patient relationships, earlier technologies had already introduced similar dynamics. Pharmaceutical testing and medical imaging are key examples.

For many years, aside from the stethoscope and visual observation, care practices lacked tools to truly investigate the internal human body. The most essential method was palpation, accompanied by the patient's medical history, varying levels of detail about the family context and, crucially, the spoken information collected during the medical visit – a defining difference between doctors and veterinarians.

Few tools were available, and diagnoses were based on relatively simple assessments. This began to change over the course of the 20th century with advances in sampling and testing technology – refined by the pharmaceutical industry – and the emergence of urban medical biology laboratories conducting substance measurements, tissue analyses and more). More recently, the pharmaceutical industry has also developed self-administered tests for general public use.

Medical testing and imaging have become increasingly central to the patient–doctor relationship, to the point of becoming dominant, often at the expense of in-person, location-based examination.

Doctors now prescribe many additional tests with specialist colleagues and/or medical biology laboratories before making a diagnosis, monitoring changes in a patient's health or even for preventive purposes. The importance of the doctor–patient relationship, and the unique nature of in-person interactions between doctors and patients, has begun to fade.

While pharmaceutical tests and medical imaging expand the traditional "face to face" meeting, shifting the highly individualized doctor–patient relationship towards a more collective one, public authorities support this shift by calling for smoother exchanges between professionals and institutions.

The family doctor, now acting as a referring physician before treatment is initiated, remains the patient's primary point of contact, the "conductor" coordinating these other interactions with various specialists.

Let us take a moment to consider the contribution of medical imaging to advancing our understanding of the human body. Tomography, which provides images of thin slices through the body, exemplifies this progress, revealing connections that are sometimes invisible in standard images.

However, in the context relevant to our discussion, the contribution of medical imaging is above all symptomatic of the instrumentalization of networking: medical imaging, an inter-disciplinary specialization that supports other specializations, (i) requires the establishment of specific connections between various actors in the treatment pathway, centered around the patient; (ii) relies on the hybridization of management data (administrative tracking of patients) and medical data (examination reports, digitized imaging files); and (iii) encourages the convergence and aggregation of multiple information systems.

The imaging file plays a key role in setting up a shared medical record. The digital health record, rolled out in 2022, bears testimony to this. Building on the computerized patient record, and later the shared record, it allows all relevant parties to store information and documents among health professionals. These professionals are encouraged to complete it.

All that remains is for patients – and also health professionals – to actually use it, which is no easy task given the repeated failures of past system implementations.

1.4. Distance and proximity: development of remote health services between the pandemic and a shortage in care provision

As we have just seen, the doctor–patient relationship has been reconfigured by the sudden emergence of medical biology and the significant contribution of medical imaging. The broad development of remote services has continued this trend, creating increasing numbers of sociotechnical mediations that symbolize a more liberal healthcare economy. These can be grouped into four main configurations:

– Remote patient monitoring is associated with the convergence between the modernization of the healthcare sector, patient-centered re-engineering and improvements in the quality of care.

– The creation of technical networking in the health industry is associated with the convergence between the circulation of information and promotion of care management methods.

– Remote appointments are associated with the substitution of the traditional physical meeting between doctor and patient in what is known as the patient–doctor relationship.

– Remote services (in the strict sense) are closely linked to the convergence between the industrialization of activities and the commercial opening of care services.

The objective: to reduce health expenditure at all costs and in various ways. There can be no doubt that digital, information and communication technologies have played a significant role because, according to Binst and Schweyer, they allowed "structures to be forgotten in order to improve the way requirements are considered in healthcare" (1995) and brought a progressive vision to the forefront by developing "cutting-edge medicine".

The available offerings are either intended for health professionals for their own use, intended for health professionals to prescribe to their patients, or addressed directly, with no intermediary, to patients.

Let us take a moment to focus on remote patient monitoring. At once a technical, organizational and informational device, it enables remote medical assistance and/or monitoring by connecting a patient with data about their health status. The data are collected using sensing, recording and automated transmission tools designed for healthcare personnel (doctors, nurses, technicians) who are in a position to analyze the patient's condition and remotely initiate appropriate urgent actions, or other interventions of various kinds.

This device, which may vary in sophistication, clearly illustrates that the way remote monitoring is discussed warrants its inclusion in the implementation of the shift to ambulatory care. In this way, it enables hospital stays to be shortened, patient travel to and between care units to be reduced, and it even allows patients to remain at home. In the latter case, it is considered capable of engaging and involving the patient and/or their close family and friends in treatments that are less costly because they are carried out outside of medical facilities – with corresponding impacts on the organization of care delivery and on the doctor–patient relationship (Carré and Lacroix 2001b).

Five key changes explain the increase in remote services and, indirectly, pave the way for the arrival of connected objects in healthcare and their future uses:

– The expansion of the legislative and regulatory framework from telemedicine to e-health, beginning in the 2010–2020 decade, reflects the government's commitment to deploying digital health in France. For example, remote medical appointments have been authorized and reimbursed by French national health insurance since September 15, 2018. This practice became significantly more common during the Covid-19 pandemic and the associated lockdowns.

– The implementation of the GDPR (granting rights of access, correction and objection to data processing) is a key factor in ensuring that data and data processing remain confidential. A contributing factor is the deployment of the French National Health Data System (SNDS).

– Advances in technology have encouraged increasingly individualized uses through the widespread adoption of smartphones and apps[5], the development of digital platforms

5 Referring to software that can be downloaded to a smartphone or tablet and accessed via an internet connection, offering specific functionality and a defined user experience. An app is also distinct in its ability to collect data.

and the emergence of connected objects – alongside the acceptance of the TCP-IP dual communication protocol for all actors in the healthcare sector.

– The Covid-19 pandemic, which required restrictions on movement and/or lockdowns to prevent transmission, forced healthcare professionals and patients to rely on digital tools for remote appointments.

– A shortage of doctors, driven by an imposed *numerus clausus* policy designed to limit the number of medical graduates, has for years reduced the availability of care and made access more difficult – especially given the uneven distribution of doctors across the country. Facing an insufficient number of physicians, disengagement by public authorities and growing patient dissatisfaction, some see remote services as a way to resolve, or at least mitigate, the shortage of general practitioners and specialists. Others, often newcomers to the healthcare field, have seized the opportunity, including GAFAM, Doctolib and various industrial or financial groups. Examples include healthcare clinics offering appointments seven days a week, managed by financial groups that had previously not been involved in the healthcare sector.

1.5. Connected objects?

Before we go any further with our analysis, it is necessary to make a quick detour to clarify what we mean by "connected objects." The goal here is not to provide a detailed analysis of the connected object market and their adoption, but rather to offer some useful information for consideration.

To draw up a general overview, let us look at some information highlighted in the 2022 report by *France Stratégie*, entitled "The World of the Internet, Objects and Dynamics to Be Mastered." What does this report show us? That the reality is complex to observe and define. That there is no globally accepted definition on a worldwide scale. That there is a notable lack of reliable statistics tools to quantify them. Estimates vary widely, and data fluctuate from one research organization to another. The number of connected objects, "according to the sources consulted in 2020, ranges from 18 billion to 78 billion" (*France Stratégie* 2022, p. 12). According to one estimate, there are 1.8 billion in Europe, including 244 million in France (ADEME and ARCEP 2022). All sources agree that the market is expected to more than double between 2020 and 2030 (AIE 2019)[6].

The main sources of divergence come from whether studies include only complex, traditional digital objects (computers, telephones, televisions, etc.) or focus solely on new categories of connected objects, such as sensors, smart meters, glasses, watches, RFID chips, voice assistants and vehicles, among others (*France Stratégie* 2022, p. 33). As a result, there are many futurological debates, recommendations and prescriptions that, with rare exceptions, reflect a bias in favor of these objects and the development of what is expected to be a highly promising future market. Some see these technologies as a form of disruptive innovation capable of creating ecosystems. Connected objects – supported by digital technology, connectivity and interactivity – are credited with improving or fostering the development of new products and services (CNUCED 2021).

In spite of all these reservations and apprehensions, we can estimate that connected objects are devices that are connected – most often by a wired or wireless connection – to the Internet, enabling data to be collected, stored, viewed, transferred and processed (interoperability). These devices

---

6 We also note the difficulty of accessing the most recent studies, which are not publicly available and may only be viewed after paying a fee. Given the amounts requested, these studies are often out of reach for researchers, particularly in the Humanities and Social Sciences.

are most often connected for the general public via a smartphone, tablet, watch or ring, along with a mobile app and a web platform through which assistance or an opinion on the collected data can be obtained, based on indicators with varying degrees of sophistication. The architecture consists of sensors, networks, software applications, private and sensitive data, and services. Connected objects have six characteristics specific to their use that facilitate their widespread adoption, the well-known 6 As: *Anything*, *Anyone*, *Anytime*, *Anyplace*, *Anyservices* and *Anynetwork*.

What about in the healthcare sector? While connected objects have developed significantly in the domestic field (home automation), in the field of the quantified self (quantification or automatic self-measurement), in industrial sectors (maintenance and logistics), and to a lesser extent in transportation and travel, the healthcare sector was still, in 2024, only marginally affected, even though experimentation has become more frequent in recent years and the first applications are beginning to appear.

In any case, in the eyes of the public authorities, they are part of a "new" direction in medicine. Namely, the so-called "5 Ps": personalized, preventive, predictive, participatory and proven.

1.6. Connected objects at the frontiers of "well-being" and "healthcare"[7]

It is useful to distinguish between two types of connected objects in healthcare:

– connected objects directly tied to a specific medical protocol and operating within a legal and legislative framework for data monitoring and oversight – a framework

7 Note that from this section onward, most of the data used in the analysis come from research carried out by Carré et al. (2021).

that, it must be said, is still not fully consolidated. These devices are used to monitor patients' health by tracking specific measurements such as blood pressure, heart rate and sleep quality, among others;

– connected objects that do not have medical device status (no medical purpose in the strict sense) and instead fall within the "well-being" category. It is worth noting that the boundary between "well-being" and "health" is often blurred, creating a "gray zone" in classification;

– a note of caution: a connected object will not do everything – its functionality depends on the applications and services it supports. Regarding health-related software, the French National Agency for Medicines and Health Products Safety (ANSM) follows the legal precedent established on December 7, 2017. According to this agency, the following devices must not be considered medical devices[8]:

- those intended only for observing the patient to ensure treatment adherence,

- those whose sole purpose is to communicate data to a doctor, without an alert function,

- those used during sports or physical training, or with functions aimed at aesthetics, comfort or improving sports performance,

- those producing results that lead only to a generic diagnosis (e.g. for a group of patients for statistical purposes or for an epidemiological field of study),

- those intended solely for administrative management such as storing or archiving data (e.g. a database or digital library containing medical information) without actively processing or using them.

8 "Software and mobile applications in health". See: https://ansm.sante.fr/documents/reference/logiciels-et-applications-mobiles-en-sante.

To make sense of this, we can look at the case of sleeptech as an example. According to a survey conducted in France by the French National Institute for Sleep and Vigilance (INSV), "36% of French people suffer from at least one sleep disorder"[9]. The global market is promising, estimated at $80 billion at the end of 2020, with a growth rate of 5.82% between 2017 and 2021[10]. This market is still emerging in Europe, whereas in North America it was already valued at $60 billion in 2018.

The market is evolving in two distinct directions: connected objects that are part of a medical protocol, and connected objects that are part of a well-being approach.

– In the first example, concerning the treatment of sleep apnea, this condition is often underdiagnosed or insufficiently addressed from a medical perspective[11] compared to other pathologies such as diabetes. It nevertheless represents a promising market. A doctor specializing in the field prescribes a monitoring protocol that includes a connected night-time ventilation device (PPC), provided by a recognized healthcare service provider, along with a software application. The data are sent and stored with an approved host. The patient is supported by the doctor and/or their team for ongoing clinical monitoring. The deployment of connected objects in healthcare is also intended to facilitate, among other things, the treatment of chronic illnesses (Simon 2017).

– In the second case, we are in a different configuration. Here, we see a new practice that is being developed: the quantified self (self-quantification or self-measurement)

---

9 See: https://www.geo.fr/voyage/le-sommeil-des-francais-en-12-chiffres-cles-186164.

10 See: https://www.businesswire.com/news/home/20171222005186/en/Global-Sleeping-Aids-Market-2017-2021-Increasing-Technological.

11 According to Pierre Escourrou, a cardiologist and sleep specialist at the Interdisciplinary Center for Sleep in Paris, training in this pathology is recent: it began in 2017.

enables individuals to act (empowerment) on their own bodies. In this case, there are no third-party prescribers; anyone can equip themselves with a connected object for a wide range of reasons – to listen to their body, prevent a heart attack, assess their heart rate, improve their sleep, monitor their physical performance (sports), or maintain their health and a healthy lifestyle. This practice contributes to greater self-awareness of our own capacities and body (Pharabod et al. 2013; Arruabarrena 2016; Lupton 2016). Informed digital advice is provided following a personalized assessment of the data sent. There are two trends: one involves reappropriating knowledge of our body to seek or return to a healthy lifestyle, and the other involves improving the body's performance potential (body-as-factory).

Thus, blood pressure monitors, pedometers, smart watches, etc., which can be used for remote monitoring for preventative purposes or to support treatment of chronic illnesses or disorders – encouraging on-site observation, measurement and assessment – become ways to implement prevention and provide assurance or reassurance. Some industrial insurance providers have understood this very well, as we will see in section 1.8.

1.7. Illustration of a "gray" zone

Let us consider for a moment a smart watch, the Apple Watch Series 4, to more closely examine the position of connected objects between "well-being" and "health." This is the first connected object not originally designed for the healthcare sector but which nevertheless has a health function recognized as a medical device. The U.S. Food and Drug Administration (FDA) was the first to recognize its capacity as a medical device for performing an ECG (electrocardiogram) via an application. In March 2019, this functionality was also recognized as a medical device in Europe.

The "ECG1" app and the atrial fibrillation (AF) feature issue an alert to users if signs of AF – the most common form of arrythmia – are detected. Untreated AF is considered one of the main causes of stroke and is the second leading cause of death worldwide. The Atrial Fibrillation Association (AF) estimates that nearly 1.5 million people in the United Kingdom alone suffer from AF, with about one-third likely unaware of it[12].

What can be said about the effectiveness of this ECG device? While it can detect certain conditions, it is worth remembering that in 2018, Dr. Leenhardt, a cardiologist at Bichat Hospital in Paris, noted: "With a cardiologist or in a hospital, electrocardiographs use twelve leads (measurement points). An electrocardiograph provided by an Apple Watch uses only one, which does not allow for detection of a myocardial infarction or other serious cardiac irregularities"[13].

The Apple Watch's ECG capability raises questions about the permeability of the boundaries between connected objects for "health" and those for "well-being". The CE mark obtained by Apple undoubtedly inspires confidence among the public and institutions in relation to e-health. Currently, we are in a "gray zone", acknowledges Pierre Trudelle[14] of the HAS. Only a "medical purpose" defines a medical device, and this purpose can be interpreted as treatment or diagnosis. This is why, according to him, a patient education tool would not require a request for medical device classification.

12 See: https://www.apple.com/fr/newsroom/2019/03/ecg-app-and-irregular-rhythm-notification-on-apple-watch-available-today-across-europe-and-hong-kong/.

13 Words gathered by Marine Benoi in her article "The Apple Watch Electrocardiogram, a Gadget or a True Medical Advance", *Sciences et avenir*, April 11, 2019.

14 Interview of Pierre Trudelle by Sarah Sandré on March 18, 2021.

1.8. Arrival of new disruptive inputs in the healthcare field

Even if developments are taking place, connected objects are much more numerous in the field of "well-being" than in the field of "health". Industrial companies have fully understood that well-being is the first step on the path to good health. This is not incorrect. Indeed, in 1946 the World Health Organization (WHO) defined health as "a complete state of physical, mental, and social well-being, and not just a lack of illness or infirmity"[15]. On this point, a trend reversal has been seen in recent years in the United States. IQVIA[16] observes that the development of applications in "health" is now dominant, to the detriment of so-called "well-being" applications.

Certain industrial pharmaceutical companies have begun to take into account the permeability of barriers between "health" and "well-being". As an example, Sanofi brought out a chatbot in 2017, i.e. a conversational agent, to support people suffering from insomnia while recommending that they take food supplements. Others, such as software developers, who had previously participated in the development of remote services in healthcare, now offer their assistance with the implementation of devices in association with connected objects. Regarding the industrial company Air Liquide, which has expertise in healthcare, it is turning towards start-up acquisitions as a business strategy. This strategy is widely used to appropriate new expertise in a key field. In 2018, Air Liquide purchased the start-up EOVE, which designed a digital device intended for use by patients suffering from respiratory distress.

15 Constitution of the World Health Organization, 45th edition, October 2006.

16 See: https://www.iqvia.com/insights/the-iqvia-institute/reports/the-growing-value-of-digital-health. The report only takes into account information from the United States of America.

We now turn to new inputs, as they demonstrate the changes occurring in the healthcare field. This is primarily the case for insurers, start-ups and industrial companies in communications and networking (GAFAM).

Insurers have been taking an interest in digital healthcare for a number of years. They have recognized the advantage of connected objects, applications, and devices for remote monitoring, remote appointments and remote services. This is why they have invested in incubators and created partnerships with various actors in the digital field. AXA was the first French insurer to offer smart bracelets to its clients. Fitbit, a bracelet that counts steps, provides an indication of an individual's daily physical activity. Starting from the principle that an insured person in good health is financially preferable to one who does not make an effort to maintain a healthy lifestyle, in return, clients of companies providing supplementary health insurance are given advantages in the form of indirect rewards for insured persons who provide their data (bonuses, reductions, free connected objects) regarding good physical performance[17]. Malakoff Médéric is not to be outdone. It provides its clients with a smart belt[18] (LUMOBack), which has a sensor that tracks the daily movements of an individual and vibrates when it detects poor posture. Combined with a smartphone app, it helps the user correct their posture and relieve back pain. This insurer was also the first to have their application on an Apple Watch. Harmonie Mutuelle has developed a connected health guide and remote medicine solutions. This insurance provider also develops partnerships with start-ups or incubators in the same way as Malakoff Médéric. In addition, we would note that remote appointments and remote medicine devices are increasingly integrated into

17 See: https://www.meilleure-innovation.com/axa-objet-connecte/.

18 See: https://www.silvereco.fr/ids-sante-et-malakoff-mederic-unis-autour-de-la-sante-connectee/3147561/.

supplementary health insurance offers[19]. The e-health services thus occupy an increasingly important position in the differentiation strategy used by different insurers. A field where they are far from being the only ones to invest, as we will now see.

Start-ups are actors who develop innovative projects, whether digital or not. The French Public Investment Bank (BPI) published a map summarizing start-ups in e-health in March 2018. There were 297 of them. In France, the most well-known start-ups in "health" among the wider public are Fitbit and Dreem, whose innovations are more closely related to "well-being" than to medical devices. By way of illustration, Dreem sells a band containing sensors and an algorithm that analyzes sleep phases. An application is associated with this connected object, offering a user support program to improve sleep quality[20]. All of this comes with a personalized coaching program.

GAFAM (Google, Apple, Facebook, Amazon, Microsoft), etc., are industrial communication and digital networking companies and, a priori, have no expertise in healthcare. However, some of them have adopted positions in a range of ways via digital devices. Health is therefore affected by the incursion of new actors from the digital world. Since more often than not, they do not have any expertise, they undeniably face difficulties understanding the rules, protocols and codes that govern organizational modes and care practices.

---

19 See: https://www.argusdelassurance.com/acteurs/e-sante-quand-les-assureurs-s-initient-a-la-telemedecine.118218 ; https://www.argusdelassurance.com/assurance-de-personnes/sante/telemedecine-la-prise-en-charge-de-la-teleconsultation-medicale-prolongee.166941.

20 See: https://www.sciencesetavenir.fr/sante/sommeil/nous-avons-teste-le-bandeau-de-sommeil-dreem-2_133955.

This is why some establish partnerships. As is the case with Google, for example, which has increased its number of collaborations with several pharmaceutical laboratories, in particular Sanofi and Novartis, and stands out due to its multiple collaborations, especially with the National Health Service (NHS) in Great Britain. This does not eliminate the question of health data sovereignty. This industrial communication company has also created its own investment fund for e-health and biotechnology: Google Ventures – a $100-million fund started in 2017[21]. Regarding Amazon, Facebook and Apple, they have attempted to make headway into the e-health market in another way via their expertise acquired in logistics: this is the case with Amazon, which is developing a medication sales platform and the delivery of pharmaceutical products. Facebook brings its expertise in advertising spaces and data collection. Apple stands out due to its smartwatch, the Apple Watch. This is an innovation for the general public consisting of a medical device, the ECG, which enables electrocardiograms to be performed, as we saw earlier, since the application has been recognized as a "medical device" on a connected object intended for the general public[22]. It is interesting to note that this watch has progressively integrated into other health applications, which have not yet been designated medical devices: oximeter (able to continuously measure the quantity of oxygen circulating in the arteries) and nightware (an application to study sleep quality). For the moment, the BATX (Baïdu, Alibaba, Tencent, Xiaomi), in other words the Chinese equivalent of GAFAM, are not yet present in the healthcare field in Europe.

Aside from GAFAM, the IT firm IBM, for example, is not to be outdone. Through IBM Watson, this large company

21 See: https://www.frenchweb.fr/le-fondateur-de-google-ventures-lance-un-fonds-de-100-millions-de-dollars-pour-la-sante/284487.

22 See: https://www.sciencesetavenir.fr/sante/coeur-et-cardio/que-vaut-vraiment-la-fonction-ecg-de-l-apple-watch_132938.

invests in the healthcare sector by developing many partnerships around the world with hospitals or companies. In 2018, IBM Watson announced a partnership with the company Guerbet[23], which specializes in medical imaging. IBM Watson also distinguishes itself from its competitors by having launched one of the main partnerships in terms of artificial intelligence (AI) between a digital company and actors in the healthcare sector.

1.9. Increase in data and data sovereignty in healthcare

The importance of industrial communication and networking companies, such as GAFAM, raises questions about dependence on globalized companies collecting, hosting and processing sensitive data and about the type of protection in place for patients' personal data. Following a decision in July 2020 by the European Court of Justice regarding the invalidation of the Privacy Shield, a legal mechanism that allowed the personal data of Europeans to be used by non-Europeans (particularly the GAFAM), the French data protection organization *Commission nationale de l'informatique et des libertés* (CNIL) issued an opinion stating that an American company would no longer be able to host a French or a European database:

> European citizens' data may no longer be entrusted to an American company, even if this company has a headquarters and servers within the European Union. This situation must result in modification of the hosting conditions of the health data platform, and the conditions of other

23 See: https://www.guerbet.com/fr/actualites/guerbet-et-ibm-watson-health-annoncent-un-partenariat-strategique-alliant-l-intelligence-artificielle-a-l-imagerie-medicale/.

> health data storage facilities hosted by companies subject to United States law[24].

Many healthcare actors using, for example, Microsoft solutions have been obliged to review the hosting conditions at the risk of being refused, by the CNIL, their "processing authorizations for these data, particularly in the context of scientific research"[25]. Despite the opinion of the CNIL, the French State Council has rejected the idea of suspending the conditions for hosting and data processing that were implemented by Microsoft, partly due to the health situation (pandemic related to Covid-19). On the other hand, it has preliminarily estimated, regarding the contract concluded between France and Microsoft[26], that no data will be transferred outside the European Union. This is important because the convergence between connected objects, artificial intelligence (AI) and processing of megadata is already drawing up what some are calling an "Internet of Behaviors" (Arruabarrena 2021–2022). Thus, connected objects, applications and services contribute to a greater datafication of the healthcare field (sensing, transmission, processing, storage and circulation of sensitive data that can be personalized). And it raises questions when "general public" connected objects nourish and feed industrial communication or networking companies but elude the healthcare sector.

---

24 Cited in the article by Alice Vitard, "Microsoft Must Remove Itself from the Health Data Hub, According to the CNIL", L'Usine Digitale, October 9, 2020. Available at: https://www.usine-digitale.fr/article/microsoft-doit-se-retirer-du-health-data-hub-d-apres-la-cnil. N1014634.

25 *Ibid.*

26 "The French State Council, the *Conseil d'État*, cites the addendum concluded in early September, which stipulates that "Microsoft will not share the data processed with the public authorities unless required to by law." Contrary to what the requesting parties claim, this exception does not only refer exclusively to American information but also to EU member states, the judge noted. To be sure of this, "it is a good idea to specify it" later, he added (*ibid.*).

1.10. Increase in sociotechnical mediations, strengthening a process of moving closer together through detachment and connected uses

The sudden rise in sociotechnical mediations in the healthcare sector is not new, as we have observed with the development of medical biology and medical imaging initially, and later during the development of remote healthcare services, encouraging new inputs into the healthcare field. In 2025, connected objects and their application software now contribute to this. The increasingly significant involvement of these sociotechnical mediations has progressively broadened the patient–doctor relationship, which in medicine refers to the very particular relationship between a doctor and their patient during an appointment in a specific location.

These mediations also have another characteristic, namely strengthening a process of moving closer together through detachment from medical practices and care. Without a doubt, we can confirm that this change, which was already in place, became highly radicalized during the Covid-19 pandemic and for several years thereafter, with the extreme difficulty that patients have faced obtaining medical appointments within a suitable time frame. The Covid-19 pandemic showcased all the opportunities that remote digital healthcare represents: online appointments, the increase in remote appointments and the multiplication of apps in e-health and in well-being.

As proof of the above, remote appointments[27] by general practitioners increased from 3,000 in 2018 to 13.5 million in

27 Remote appointments comprise four different sociotechnical configurations: with the family doctor via a video appointment; an emergency appointment with a doctor that the patient has sought out, for example, on Doctolib; the use of a specialist digital platform offering an available doctor, as is the case on the Médecin Direct platform; and remote appointments from a station installed inside a pharmacy or in a shopping center with a doctor chosen for you. Some operators, such as Qare, offer

2020 (French Republic and DRESS, 2022). Even though this is considerable progress, it is important to specify that these appointments represented less than 6% of the activity of general practitioners.

Public authorities, such as the medical corps, also see in this a way to facilitate access to care for patients. In this context of the government's "acceleration strategy" for digital health (e-health), in 2021, the public authorities rolled out an incentive-based policy to encourage the deployment of connected objects in healthcare (see the France 2030 roadmap).

No doubt, the process of digitalization has contributed to facilitating the movement of patients closer to care personnel, while moving them further away from care structures (hospitals, clinics, grassroots medical services).

This renewed interest in remote healthcare, which, according to information, is as close as possible to the patients, also plays a role, it must be said, in reducing healthcare costs by integrating digital solutions to seek a better service and, above all, greater productivity[28]. Here, we again see one of the dominant elements that guided the implementation of the sociosanitary reform at the turn of the 1990s.

Undeniably, connected objects in "healthcare" will hold a more important position from now on. Their novelty now resides:

appointments seven days a week, from very early in the morning until midnight.

28 Words contributed by Charha Louafi, director of the Fonds Patient Autonome, "BPI France Strengthens Its Support for the Healthcare Sector", BPI France website, 6 April 2021. See: https://www.bpifrance.fr/A-la-une/Actualites/Bpifrance-renforce-son-soutien-au-secteur-de-la-sante-52034.

– on the one hand, in the experimentation of care devices that are still tentative, implemented by healthcare personnel, which often requires the provision of real support for the patient in using digital technologies and adhering to protocols that must be followed. This involves developing patient education. Indeed, use studies show that patients are too often left to their own devices and their own digital culture, without taking into account the inequalities in technical skills and access to their data (Carré et al. 2021)[29];

– on the other hand, the installation of a flood wave overwhelming care practices already established within the healthcare sector places connected objects within a desired or imposed societal change, namely the quantified self (quantification or self-measurement of ourselves), practices which are implemented by using connected objects intended for the general public, which cannot be considered devices with a medical purpose, in order to implement coaching, well-being or self-measurement practices in order to train to improve lifestyle or boost good conscience. In this case, self-tracking favors monitoring of the user over a long time period. Certain analysts see a form of "managerialization of oneself" (Dagiral et al. 2019), others a "hygienization" or "medicalization of daily life" (Calvignac 2021). Thus, referring back to the deployment of social norms in terms of healthcare and well-being, the implementation of which is largely facilitated by resorting to connections arising from other information practices on social media networks, as well as online purchasing, remote bank management, etc. The objective of new arrivals in the healthcare field between "well-being" and "health" is to make any individual into a potential client and not a patient, thus marginalizing the medical practices instituted. This disruption is only made possible by implementing digital tools, as has already been the case in other fields or activity sectors, as pointed out earlier in the text.

29 The use studies were carried out by Geneviève Vidal.

1.11. References

ADEME and ARCEP (2022). Évaluation de l'impact environnemental du numérique France et analyse prospective. Summary note, January.

AIE (2019). Total energy model for connected devices. Programme de coopération technique, Agence internationale de l'énergie.

Arruabarrena, B. (2016). Le Soi augmenté : les pratiques numériques de quantification de soi comme dispositif de médiation pour l'action. PhD Thesis, Conservatoire national des arts et métiers, Paris.

Arruabarrena, B. (2021–2022). Objets connectés : penser les enjeux des technologies connectées sous l'angle de la médiation infocommunicationnelle. *Tic & Société*, 15(2–3). doi: 10.4000/ticetsociete.6207.

Binst, M. and Schweyer, F.-X. (1995). *La santé otage de son système*. InterEditions, Paris.

Calvignac, C. (2021). Le design des dispositifs d'automesure du sommeil. *Tic & Société*, 15(2–3). doi: 10.4000/ticetsociete.6214.

Carré, D. (2001). À la frontière du marchand et du non marchand. Place des téléservices dans la réorganisation du secteur de la santé. *Études de communication*, 23, 31–49.

Carré, D. and Lacroix, J.-G. (1999). Virage ambulatoire et autoroutes de l'information dans le secteur de la santé. *Sciences de la société*, 47, 23–41.

Carré, D. and Lacroix, J.-G. (eds) (2001a). La santé et les autoroutes de l'information. In *La greffe informatique*. L'Harmattan, Paris.

Carré, D. and Lacroix, J.-G. (2001b). Incidences de l'usage du télémonitoring sur l'organisation du travail médical et la relation médecin-patient, e-usages. *Les usages et les services de télécommunications (ICUST)*. ENST, Paris, 470–481.

Carré, D. and Panico, R. (1998). Imagerie médicale, rationalisation et mise en réseau du soin. Vers une recomposition des territoires, des organisations et des pratiques professionnelles dans le secteur de la santé. *Médiations sociales, système d'information et réseaux de communication*, Metz, 141–149.

Carré, D. and Panico, R. (2003). Informationnalisation en santé ou le rapprochement par l'éloignement ? Éléments d'analyse d'une gestion économique et organisationnelle de la distance socio-sanitaire. In *Bogues. Globalisme et pluralisme*, Lacroix, J.-G. and Tremblay, G. (eds). Presses de l'Université de Laval, Quebec, 101–116.

Carré, D., Vidal G., Sandré S. (2021). Recherche sur les objets connectés en santé (ROCS-SHS). Research Report, AP-HP, Paris.

CNUCED (2021). Technology and innovations. Catching technological waves: Innovation with equity. Report [Online]. Available at: https://unctad.org/fr/isar/page/technology-and-innovation-report-2021.

Dagiral, E., Licoppe, C., Martin, O., Pharabod, A.-S. (2019). Le Quantified Self en question(s) : un état des lieux des travaux en sciences sociales consacrés à l'automesure des individus. *Réseaux*, 2019/4(216), 17–54.

DREES (2022). Sept téléconsultations de médecine générale sur dix concernent des patients des grands pôle urbains en 2021. *Études et résultats*, 1249.

France stratégie (2022) Le monde de l'Internet des objets des dynamiques à maîtriser. Report, République française [Online]. Available at: https://www.strategie.gouv.fr/publications/monde-de-linternet-objets-dynamiques-maitriser.

Julien, P.A., Vermot-Desroches, B., Chouinard, J.Y. (1987). La santé, les services sociaux et les personnes âgées en 2001 au Québec. Rapport technique auprès du ministère de la Santé et des Services sociaux, Université du Québec à Trois-Rivières.

Lupton, D. (2016). *The Quantified Self*. Polity Press, Cambridge.

Morelle, A. (1997). Pour une politique de santé publique. *Esprit*, 24–33.

Pharabod, A.-S., Nikolski, V., Granjon, F. (2013). La mise en chiffres de soi. *Réseaux*, 1, 97–129. doi: 10.3917/res.177.0097.

Simon, P. (2017). Les leçons apprises des principales études sur les objets connectés en télémédecine et santé mobile. *European Research in Telemedecin/La Recherche européenne en télémédecine*, 6(2), 67–77. doi: 10.1016/j.eurtel.2017.06.003.

2

Connected Sleep: From Individualizing to Modeling the Doctor–Patient Relationship

With this contribution, we aim to outline the issues associated with the automation and the use of connected objects in medicine, in order to study their impacts on medical practices, relying on the interviews of 20 doctors treating patients affected by sleep apnea syndrome (SAS)[1]. Patients were equipped with a medical device that applied continuous positive airway pressure (CPAP) and recorded data during sleep. These data were then collected by the service providers and made available to both doctors and patients who could consult the data via an app on their smartphone. The challenge of this study was to identify the doctors' use of connected devices and the repercussions on the relationship

Chapter written by Christian PAPILLOUD, Geneviève VIDAL and Yanita ANDONOVA.

1 The data from the interviews were collected in 2021 in the context of the French project entitled "Challenges and Effects of Digital Devices on the Medical Relationship with the Treatment Prism for Sleep Apnea Syndrome (During Covid-19)" [*Enjeux et effets des dispositifs numériques sur la relation médicale au prisme du traitement du syndrome des apnées du sommeil (en période de Covid-19)*], coordinated by Cristina Lindenmeyer (2022). We want to thank all researchers involved in the initial project and Cécilia Jauniau for her work in collecting the data.

with their patients[2]. The purpose was not to interview patients, but to focus solely on the doctors' perceptions and representations. The interviews were structured around three main themes: the practices and uses of connected objects by doctors, the relationships between doctors and patients within this context and the challenges posed by connected objects for medical practices. Each theme was introduced with a question that invited the responder to develop a narrative, which led to further questions aimed at clarifying their arguments. The 20 interviews were recorded and transcribed, thus constituting the corpus of answers on which this study is based.

This study was performed using the method of topic modeling for cases within the context of medicine (Papilloud and Schultze 2023). For the doctors interviewed, this connected object – the CPAP device – operates a series of movements that affect the doctor–patient relationship as follows: the more prevalent the technology is to the doctor, the more it is elected as the preferred agent, which maintains and improves the relationship, allowing for the incorporation of additional complementary technologies. We discuss this result by suggesting a turning point in medical practices: the importance of the patient to medicine is equal to the data that they produce on this pathology, the complexities of which the doctor hopes to reveal, while monitoring the multiple changes in time and space. Before we discuss our

2 The investigation was carried out with doctors who answered the call for participants issued by Marie-Pia D'Ortho sent out via the Hypnos network of doctors. See http://reseau-hypnos.org/p/hypnos.htm; see also (Carré et al. 2021). Given the complexities of the pandemic, the interviews, which averaged 60 minutes in length, were conducted remotely. The doctors also completed an online self-administered questionnaire about their medical practice in the context of monitoring patients with sleep apnea, equipped with a connected machine.

analysis, we will provide brief details of the methodology used to analyze the interviews.

2.1. Methodology

Preparing our corpus for topic modeling analysis required transforming the raw data into a format that could be used for modeling. First, we converted the documents into text format (*.txt) and we anonymized the respondents (identified by the letter E followed by a number for each individual).

The interviews were conducted according to a structured guide that focused on the three general themes mentioned previously. We extracted these themes from the interviews by labeling each series of corresponding questions as A (practices and uses), B (doctor–patient relationships) and C (automation issues).

To ensure the arguments presented are legible, we decided to remove the investigators' contributions from the interviews, so that only the respondents' answers are visible. These responses provide information on the way the interviewee filled in the category to which the series of questions was related. We consider, for example, that all the answers to questions under category A (practices and uses) represent the various ways in which doctors develop a discourse on how they represent medical practice and the use of digital devices and connected objects in their professional activity.

Then, we grouped the answers related to each of the three categories in the interview guide together into separate files (A, B, C). We then proceeded to model them. These data were modeled using the MTA software (see: https://github.com/cp1972/mta-app), which offers a modeling process based on algorithms, enabling similar groups of themes across the entire corpus to be identified and analyzed, i.e. the topics.

These could then be linked with the doctors who addressed these themes in the interviews.

We have compared the answers based on the themes discussed in the interviews to differentiate them. Three groups of doctors have been identified: a) respondents working as self-employed practitioners, b) respondents in public hospitals and c) respondents in university hospitals.

We now propose examining the three categories and corresponding topics, each in turn, as follows:

– category A: practices and uses of digital devices and connected objects;

– category B: relationships between doctors and patients mediated by technology;

– category C: issues relating to automation in the doctor–patient relationship.

2.2. Category A: practices and uses of digital devices and connected objects

Within category A, our optimal model is composed of four topics, namely, four principles that structure the opinions of all respondents in this category, with a correlation index of 83% between these topics and all the interviews. Hence, we have made a distinction between:

– The first topic relates to the data accessible to the respondents regarding the monitoring of patients treated in the units where they work. Access to the data is a theme discussed in several contexts (related, for example, to the relationship with the service provider, to training in technology, to the legal aspects arising from the use of technology, to patient monitoring through this technology or even to appointments).

– The second topic refers to the various control operations that the respondents may perform by way of the digital technology used (verification of the technology or of the patient).

– The third topic concerns the effectiveness of the technology used, enabling rapid improvement in the observation and adjustment of patient treatment. Remote monitoring enters into play here as a decision-making tool to improve the patient's treatment.

– The fourth topic is dedicated to the development of digital technologies (CPAP devices, connected objects and AI) in the medical field, particularly for patient diagnosis and monitoring. Respondents often mentioned artificial intelligence (AI) in their discussion, on the one hand, to note the contribution made by this technology to medical practices, and, on the other hand, to highlight the questions it raises regarding the relationship with data rather than with the patient.

In what follows, we connect the results of our modeling to the discussions collected, beginning with the first theme.

2.2.1. *Data as a tool for legitimization*

A majority of the responding doctors, E1, E2, E5, E6, E14 and E17, support the theme relating to the access and accessibility of data by the service provider who administers the technology used. Among these six doctors, E5 and E17 exclusively mention this data-related theme within the context of remote patient monitoring. E5 says that she was "not born with the Internet":

> [I] grew up with the Internet, but I'm not at ease with digital technologies. [...] For me, I clearly need someone to train me first, to show me how

> the software works, how to open it, how it can be formatted, how it can be set up.

As a consequence, E5 does not access monitoring data for her patients directly, but does so through the intermediary of the service provider:

> In fact, I would ask a secretary to contact the service providers to ask them to send the machine records so that when the time came for the remote appointment, I could have the machine records with me and call the patient to discuss the results with him/her.

While in her department "everyone feels more or less the same as me", other practitioners, such as E17, feel "at ease":

> [On] the one hand, because I am particularly comfortable with everything digital, I'm very interested in it, so I have a certain ease with it. And, in addition, I also participate in the development of these platforms, indirectly, directly.

In E17's view, technology enables a daily check to be performed, and consequently, monitoring the patient's health can be carried out more precisely. The four other doctors, E1, E2, E6 and E14, differentiate themselves from the first two. Although they are primarily involved in the theme of data access, their discussion includes nonetheless dimensions relating to other topics. According to E2, for example:

> [P]reviously I only looked at the machine's SD card [memory card], which also gives me more reliable information, but since there has been this possibility to access data related to remote monitoring, I immediately look at their

> monitoring, what the main treatment criteria are, and if there is a problem, I will look at the card if they have brought it.

The service provider E2 is often a private one that installs its own remote monitoring technology. In E6's view, this often functions better than public assistance software: "I had to use the public assistance remote monitoring systems, which don't work well at all. [...] They didn't work at all, so, we didn't use them." Conversely, "I use AirView every day. [...] I find that AirView and the service provider websites are very user-friendly." This theme also arises in the discussions with E2, who would like other functions to be added to the technology so that he can be promptly informed of any changes in his patients' health. E1 also highlighted this dimension of the future development of the technology.

Lastly, doctor E14 stands apart from the others in that she includes the theme of the third topic, i.e. the effectiveness of the monitoring technology, which she did indeed appreciate, in particular during the Covid pandemic:

> [T]his remote medicine, in our particular case of monitoring via machines, really is the apotheosis, at the very limit. It really is ideal.

She continues:

> There is some interesting information that I had put in my summary at the sleep conference about RESMED [Respiratory Medicine] data, the very subject we are talking about, where they have looked at this period of lockdown in Europe, [...] And, so, we realized that in France, the number of patients under remote monitoring had not really increased very much [...]. On the other hand, the daily number of logins to the manufacturers' platforms had increased

> significantly, by about 4,000, 5,000 per day, unless I am mistaken.

In E14's view, this increase in the use of remote monitoring and remote appointments during the pandemic had dual benefits: "it enabled me to continue holding appointments almost to the same level that I would have been able to do with face-to-face appointments", while all the time avoiding "spreading the virus any further" and with no detrimental effects on the patients. Such that E14 concludes: "It [the technology] is sufficient in order for me to make entirely safe decisions today."

2.2.2. *Control as a way of guaranteeing legitimacy*

In terms of the monitoring technology, the respondents E4, E19 and E20 support this theme of verification as a guarantee of legitimacy almost exclusively. E4 states that he is comfortable: "I consider that I am the one who is making the decisions depending on the data I have at my disposal." Nevertheless, when it is used, it is presumed that there will be a continuous exchange with the service provider, which provides the technology – in the case of E4, he is in contact with several service providers – so that he can be sure that, from the technology to the production of the data, everything operates without any hitches:

> The prescription, the choice of the machine, the choice of settings, even putting a humidifier in place, etc., this is the doctor's job all the same. Then, the service provider calls me sometimes on my mobile phone, sometimes, when they are at home or when there is a problem, to ask if I want to take any action for a particular problem that has arisen.

E19 admits: "This is clearly IT, but me, I'm a doctor, I'm not a computer specialist [...] Then, there are some who love it [...] As far as I am concerned, I'm not one of them, not at all." Therefore, in E19's opinion, the challenge posed by the technology is above all related to the way it contributes to improving the patients' health:

> These are the parameters I need for the patient, to see how he feels, and clinically I question him, and I look to see if it is coherent with the results that I have, and then, if things don't add up, I ask for the memory card to investigate further and see what is happening.

In E19's view, these verification tasks are necessary because the technology sometimes makes mistakes: "The machine does its thing, and we are here to check, to look at it with a critical eye, we can't content ourselves with the conclusion made by a machine." E20 expresses his difficulty in using the technology in a more resolute manner than E19, particularly due to the huge amount of data generated: "The disadvantage is that there is too much information." According to E20, this large quantity of data leads to the use of additional data as well as patient opinions:

> When I have a lot of numbers, when I see these data, I don't say to myself, oh right, so, it's obvious, I'll do that, and there will necessarily be an improvement in my patient, because it's not always the case, so there are other data that I am missing [...] I show the screen to the patient, what I can see, I show to them.

According to E20, the fact that doctors need to go a step further than the data provided by the technology, including the patient's perspective, refers back to the doctor's obligation to do their work well using this technology. Doctors E10 and E16 also mention the themes of data access, the accessibility of data for practitioners and the effectiveness of the technology. E10 mentions difficulties with "juggling between the various service provider websites and the different software programs used by different machines", which results in lost time for him and for his colleagues. According to E16, the problem is similar, even if it presents itself less often in data collection than in the installation of the technology that he would like to have on his computer:

> The little aspect where an issue arises, is the interfacing with software programs used in the profession. In other words, I don't have them [connected medical devices] on my professional software, so, they are on the manufacturer's website, that is, the service provider's website.

Regarding the theme of the technology's effectiveness, E10 highlights its practical aspect, the fact that it makes the contact with the patients simpler – in other words, that the patient does not have to travel to the appointment – and easier to track and more reliable:

> It really is a huge convenience [...] We have access to all the data instantaneously, it takes just a second, it allows us to compare the patient's clinical feelings with the data we have.

For E10, the technology *does* have its limits, and it is not a case of making it the focal center of the care process, but "it always brings something positive". According to E16, the monitoring technology is merely an additional step in the progress made in collecting data about patients, a process that is now largely paperless. Respondent E13 is the only one to tackle the question of automated data management:

> I am not sure that there is a lot of automated processing behind it. Because when we see the

> service providers' websites which are more or less awful [...] They just look nice and are user-friendly.

Similarly, regarding the possibility of expanding the technology to artificial intelligence, E13 states:

> I'm not sure we have a lot of data to process via artificial intelligence for CPAP [...]. The service providers don't have a lot of money to invest into their websites.

In E13's eyes, the horizon for developing the monitoring technology is not something that worries him, because, for him, service providers are the ones who must take the initiative in developing this.

2.2.3. *Effectiveness of the technology*

The question of technology effectiveness was discussed particularly by the respondents E3, E7, E8, E9, E11, E12 and E18. In E18's view, "the reliability [of the monitoring technology] is great. I would give it a ten out of ten", namely because, on the one hand "it is accessible all the time, 24/7", and, on the other hand, "it greatly simplifies medical files, we don't need to print things out anymore, we don't need to receive documents... It considerably limits the number of emails, pieces of paper".

Other than the time saved and the ecological aspect, E18 associates the effectiveness of the monitoring technology with an improvement in patient treatment: "If it is a tool to reduce my workload... that's not really what I am looking for. However, if it will help me treat my patient better, then yes." As for E3, the technology "has only advantages", because "it enables a very high degree of reactivity to requests made by the patient and the service provider".

Moreover, given that the "reliability is good" and that a large majority of patients are adopting the monitoring technology, E3 does not see himself "taking a step backwards". E8, E12 and to a lesser extent E9 and E11 consider the effectiveness of the monitoring technology in relation to the verification operations on the data produced by the machine. E8 is the most emblematic respondent of this trend, because he is developing a research program based on the quality of the data provided by the remote monitoring machines:

> With regard to everything to do with observance [...] they are objects that are very reliable if you have a very high apnea-hypopnea index [...] Now, these are connected objects that have the advantage of repeated measurements, and sometimes, the diagnosis is the repetition of the measurement that enables us to consider a diagnosis.

In E8's view, the verification of the data produced by the CPAP device should make the practitioner more attentive to the use he makes of it:

> Then again, choosing between being supervised and unsupervised, I believe that today we are at a stage of the relationship with connected objects which, in my view, requires the option of supervision. And, that we should say, "I'm looking for that, please develop a learning process which shows me that.

According to E12, in this aspect of supervising the monitoring technology, it is assumed that the practitioner will undertake to verify the data:

> Total automation is not possible in this field, in my opinion, currently. Because it is not possible

> to imagine [...], that if we see apnea occur in a patient, an algorithm will immediately generate modifications in the machine settings, without human intervention. There must be verification by a human.

E11 highlights the exact requirement for the practitioner to verify the data:

> All polygraph and polysomnography reading software programs perform automatic readings. The automatic readings in question are systematically reviewed, and I make a lot of corrections.

Regarding E9, for whom this theme of data verification is less apparent, he also expresses the need for human intervention to lend meaning to the data and verify that they accurately reflect the patient's health condition.

2.2.4. *The horizon of developing technology*

This last aspect of the technology's development horizon strikingly appears in the interviews with doctors E1, E2, E6, E7, E11, E13 and E14. It is also the central theme for respondent E15. This doctor states that she is even more at ease with the monitoring technology since she "has been one of the first to use it, and even to create it". In E15's view, the technology can be further improved to play a greater role in patient monitoring and diagnostic devices. From this point of view, E15 is expecting the use of artificial intelligence to rise:

> [It will do] many things [...] much better than me. Like finding out [...] the contraindications in the combinations of medicines, the fact that, if all the clinical data could really be entered, I believe that artificial intelligence would probably be more

> efficient than I am in terms of establishing the various diagnostic hypotheses.

Indeed, AI is not intended to replace a doctor for E15, "it's just an additional tool". Nevertheless, it describes the horizon of an even greater expansion of monitoring technology, which is likely to be able to include more people and apply to different registers of pathologies.

2.3. Category B: doctor–patient relationships mediated by technology

In the context of category B, the analysis yields an optimal model with four topics and a correlation index of 92% between the four topics and all the interviews. Among these four topics, it is possible to make the following distinctions:

– The first general topic relates to the theme of the relationship between patients and doctors through the intermediary of monitoring technology and the data it generates. The particularity of this topic is that it evokes all the themes that are subsequently found in the three other topics detailed below.

– The second topic targets the theme of reliance on remote monitoring and the data generated by this monitoring, particularly with respect to the pedagogical effort made by the respondents to make their patients understand how the technology works and how it can help them.

– The third topic concerns the network environment which develops around remote monitoring, in particular towards other applications proposed by this technology, and towards other technologies (connected watches, pedometers or even social media networks).

– The fourth topic concerns the relationship with service providers, particularly with technicians who work for these service providers and are in contact with patients and doctors.

In the following, we provide a detailed examination of the opinions expressed by doctors in support of the four-topic model outlined above.

2.3.1. *The doctor–patient relationship*

Doctor E7 is emblematic of this first theme:

> Trust in the technology remains. [...] There are pneumologists who will tell you that they have many refusals regarding remote monitoring, well, this is strange because, as for me, I have never had any. Except in exceptional cases [...]. Because, if the doctor himself trusts the technology, the patient will too.

In this trust-building relationship, E7 believes that it is essential to be genuine and to inform the patient that the technology is not perfect, that it also has its limitations: "If we present the advantages and limitations honestly, and we ourselves seem to trust it, it will work." This theme of reliance also arises in the discussions with doctors E3, E4, E5 and E11. For example, E11 confides that the use of a CPAP machine "does not at all alter, I believe, the trust relationship". E4 presents similar arguments, stating:

> From the moment when we rely on these devices without losing sight of the fact that what counts is in fact the relationship and the way we listen to the patient, I don't think it affects the relationship.

In this context, E4 relies on the dialogue with the patient and on listening to the patient to improve acceptance of the monitoring technology. E3 and E5 are specific in that they delegate part of this relational aspect to service providers. In the case of E5, this is because "I have around ten patients in the office, and really, this is very few for the moment"; in the case of E3, this is because "I quite simply don't have the time to do that, to take things in hand." Like E2, E8, E13, E18 and E20, they emphasize the world of apps that are directly offered by monitoring technology, or the deployment and use of which are indirectly encouraged among patients. E18 refers, for example, to certain patients who equip themselves with connected watches, which provide information to them when they experience periods of apnea, or specific digital devices that collect data about patients who have undergone transplants. As for E13, he states: "Now, people will do an electro for you every two seconds, on their own, using their watches." In E20's view, "nowadays, with these new connected objects, some can perform diagnoses, made precisely by these objects [...] without analysis", hence the importance of dialogue with patients so that they understand the meaning of these diagnoses. Similarly, E2 comments:

> There are people who are very passionate about technology and look every day at the signals that their machines send to them, and who then report that information to me, and who even ask me if they can have the card reading software to look at themselves.

Nevertheless, this is not the case for all patients, some of whom feel observed, even under surveillance: "There are some who have said no, no, I refuse remote monitoring." However, E2 is surprised to see no adverse reactions to what she calls "policing": "there is an intrusion, but I haven't had any really negative reactions to that", an observation also made by the other respondents for whom the doctor–patient relationship and trust in the doctor and his expertise are not really jeopardized by this technology. Additionally, the device is "well protected by all data protection laws", and the service provider is required to obtain the patients' consent. From both E8's and E2's points of view, and in the eyes of the other respondents who, for the majority, supported this theme in their discussions, "the way the connected object is presented is what's most important", otherwise, patients "believe in a watch more than they believe in doctors":

> A connected object already replaces the opinion of Peter and Paul, of a cousin, or so-and-so's opinion, etc. And, this, the connected object, is something that represents the truth.

Hence, the importance that E8 attributes to the pedagogical aspect, to the way attention is paid to the patient, so as not to damage the relationship between the doctor and his patient. It is a case of not leaving the patient alone with these technologies: "We need to be very prudent about the way we express ourselves, and pay close attention to what the patient will say."

2.3.2. *Trust*

According to E6, E12, E15 and E19, trust is not only interpreted as the trust that the doctor and patient must have in their relationship, but also as the trust in the monitoring technology and, by extension, in the service providers who manage the CPAP machine from a technical standpoint. In E19's view, it is essential not to have blind trust in the technological tool:

> Either we are doctors, or we are not [...] in the end, as far as the course of action is concerned, it's the doctor who decides, it's not the machine that will say "yes, this and that needs to be done."

Comparably, E15 specifies:

> They all come with connected watches now, and they say "you see, there, I don't have deep sleep anymore, so I don't sleep at all". And this requires us to put it back into context.

The fact that monitoring technology is an aid and that its reliability should neither be overestimated, nor should blind trust be held in other apps or connected objects to which the patient has access, is in E15's view, a question of trust placed in it:

> When I choose a service provider, it's one who treats patients well and where the relations with service providers are good [...] So, in general, I only work with service providers I trust in.

In E12's view, while "a great majority [of patients] have trust in doctors, in service providers, in the team who monitors them, saying, well, 'you use the data, I have very well understood that it is useful, it's secure, etc.'", time must even so be invested in the patients, so that they correctly understand how remote monitoring works: "Time must be taken to explain to patients that these devices are not going to interfere with their health." E6 wholeheartedly agrees:

> Any patient who is equipped with the machine and who doesn't perfectly understand how it works is worried [...]. For us, as doctors, our responsibility is to reassure the patient, to explain clearly and simply how it works.

For these doctors, it is a matter of ensuring that patients trust in the technology, while helping them make reasonable use of it, particularly when patients interact with others on social media channels.

2.3.3. *Reticular extension of the doctor–patient relationship and involvement of the patient in their own treatment*

For E10, E16 and E17, it is not necessarily unwelcome for a patient to seek more information and data about their health state, including by communicating on social media networks. In E16's opinion, patients want to find out more about their health: "They ask me for their data, I show them their data, I give them a copy of their data, I provide explanations to them." While this encourages relative autonomy of patients regarding their doctor, it nevertheless must not substitute for the doctor–patient relationship. There is a "fear of being entirely replaced, at some point, in our profession", confides E10, which is why it is important to ensure the patient understands that the technology is not an end in itself, but simply "an aid":

> Then again, yes, it changes things, because, necessarily, the content of appointments is not the same for patients who know, who have more information about the illness, about treatment options.

The fact that the relationship changes as a result of greater patient involvement, as they become better informed about the monitoring technology they are using, "is not a bad thing. After all, this creates a group dynamic. Once again, whenever people are interested in their health, it's great". E17 states that this is also

> a requirement. The doctor cannot be content with simply reading a PDF if the patient has online data, ultra-detailed data, things like that, whilst it's an additional requirement. The profession is evolving.

The doctor–patient relationship is changing, but this is not the case for all patients, just "a marginal group, always geeks". Through other technologies that patients equip themselves with, such as smart watches, which often recur as examples in the respondents' discussions, or via the intermediary of discussion forums or social media networks, the reticular extension of the doctor–patient relationship enables knowledge to be shared. However, as specified by E17, "sharing data is not the same as sharing knowledge". The increase in information on the part of the patient does not, nonetheless, create an expert, but it increases the need to take what the patient says seriously to "verify", "explain" and "educate". In this context, the reliability of the technology assumes that the service providers and their technicians, who are responsible for verifying that everything works correctly for the patient, are competent. This last theme is at the heart of the discussions with E9 and E14.

2.3.4. *Service providers and their technicians*

Doctor E9 reminds us of the importance of listening to patients considering that they now have more knowledge about monitoring technology and its processing, particularly due to connected objects and the information available on social media networks. The doctor also emphasizes that in Europe, there is a legal framework that governs the sensitive nature of health data. Nevertheless, fears remain:

> The fears are above all [...] about what people are trying to sell us, in inverted commas. And, when I say that, I am thinking as much of industrial lobbies or large companies, as about what drives us to use legal proceedings.

However, E14 adds a nuance: "I think I am a little too demanding, but I find that the level of service providers, precisely since this remote monitoring has been set up, has increased, and improved." While this does not necessarily concern all service providers, it is at least the case for service providers with whom E14 works.

The challenge remains, nevertheless, to make patients accept periodic visits from service providers. While some patients would like to be visited more often, others consider these visits to be an intrusion:

> This may be, for example, technicians who come and who are going to enter the patient's room, who will sit on the bed to look at the machine.

Sometimes, the service provider does not take action quickly enough when technical issues arise, and sometimes they do not fully understand what the patient is communicating to them. This additional extension of the doctor–patient relationship must also be discussed. This task also falls under the responsibility of the doctor, who serves as the spokesperson for their patients with service providers and monitors to ensure that services are provided to these patients in the best possible way.

2.4. Category C: challenges of automation in the doctor–patient relationship

Our analysis of this last category of answers yields an optimal model with four topics, and a correlation index of 88% among them and across all interviews. This last series of questions builds on themes that we had already identified in the previous results drawn from the first two. The meaning of the four topics is derived from this, with:

– the first topic where the respondents return to the importance of the relationship with the patient in a world where different technologies increasingly cover this relationship;

– the second topic regarding the technologies that the respondents currently have and the way they can be changed, in particular by integrating other connected objects or through artificial intelligence;

– the third topic concerning the way the technology attributes responsibility to the respondents, and how this question of allocation of responsibility is involved in the doctor–patient relationship;

– the last topic regarding the influence of technology on the respondents' practice of medicine and on the daily life of their patients.

We will now provide detailed explanations of these topics, returning to our previous discussions with doctors in the order of the topics.

2.4.1. *The doctor–patient relationship faced with a sharp increase in technological devices*

Respondents E19 and E9 are among those who exclusively support this theme in their discussion. In E19's view, "time can be saved with all these tools. But I personally don't want to lose the human aspect". However, E19 *does* keep up to date with technological developments – "it's true, yes, we discuss the developments and everything" – because in her opinion, despite the disadvantages associated with these machines (E19 cites in particular the energy expenditure of machines or the surveillance performed by the national health services on patients to ensure that the machine is used correctly), "the assistance this contributes, perhaps not in any and all fields, but here, I'm in this field and I know this assists me". E9 expresses a similar relative ambivalence regarding treatments which assume that remote monitoring technologies will be used: "There is a certain time dedicated to technical aspects which has increased with respect to what it was previously [...] Then again, this has benefits, but in any case we don't have a choice." Just like E19, E9 underscores the ecological challenge of deploying technologies:

> It's an ecological catastrophe! At the same time, all the components of the processors are rare earths, rare materials, etc., that we are in the process of ransacking the planet to obtain them.

Nevertheless, E9 must stay up to date with changes in these technologies, because "things change quite quickly" and patients must be able to benefit from the advantages of the technology. In the eyes of E7, E15 and E16, whose points of view focus almost entirely on this first topic, opinions are less ambivalent. E7 says that she has never known anything other than care by remote monitoring and that, more generally, she has always practiced her profession in a technologized environment: "I am from a generation who has not known, for example, scanners printed onto boards, and who has always seen them on a computer." Therefore, for E7, technology is something familiar, something "practical", that she believes she cannot do without, because "it would not be good for the patient". Monitoring technology does not represent a significant financial cost. As for its ecological cost:

> It doesn't represent a large flow of data, contrary to what we might think. [...] It costs less than Facebook in terms of electricity. And, since the visit by the technician is more limited, the ecological cost is therefore also more limited.

E7 also views the increase in the use of connected objects for remote monitoring in a good light, all the more so given that patients "also understand what we can offer them as a digital tool, dietary treatment, sleep diary, this kind of thing". E15 follows a similar line as E7, while underlining that:

> I find the remote monitoring of CPAPs interesting but really, it's not that innovating either. But if we imagine systems that are a little more sophisticated, that will depend [...] on the day they reach maturity, then yes, they should be used.

In E15's view, the ideal would be "systems that are able to change [...] thanks to artificial intelligence [...] which learns by itself" in order to be able to "compile health-related data", in other words, to aggregate this data to divide it up more easily. According to E16, who in the past "used machines all the time", monitoring technology does not "present a problem". The only problem comes "from the lack of policies", which leads to a lack of investment in training doctors in these technologies.

A similar discussion occurred with the respondents E1, E3, E5, E8 and E14. In their view, the increase in technological devices attributes responsibility to the doctor, who must have "safeguards" (E8) that explain, "and we must explain with concrete examples relating to them" (E14), to elicit the patient's compliance with the use of these technologies. "It's a transformation, it attributes a lot of responsibility to us" (E1). It is important not to "miss the train. I think it's an irreversible change, and it is in our interest to use this type of thing as much as possible" (E3). In the case of E5, the technology generates an attribution of responsibility to the patients about their own health condition:

> I have the example of an elderly lady at the office who is a young elderly woman, who must be 68, 70 years old, you see, and who is in very, very good shape and who has had an apparatus for many years, and who knows her machine very well [...] she manages to get the number, the residual AHI which is displayed. And in fact, she manages to control the small buttons, things

> like that, and look at the information given by the machine, and so, it's true that she comes to the appointment saying to me, "look doctor, the machine is telling me this, etc., etc." I believe that this attributes responsibility in fact.

2.4.2. *The evolution of available technology*

The respondents E2, E10, E11, E12 and E20 had points of view mainly focused on this theme in their discussions. E20 is different in that a very different line of discussion is presented, even opposed to those of other respondents:

> I am oppressed. I don't manage anything. I don't manage anything at all. It's not nice, it's unpleasant, yes, this relation to time… when I finish an appointment, I haven't finished, precisely because, even if I haven't been able to look at all the platforms, I still need to look at the platforms.

In the case of E20, technology is a stress factor that can lead to patient abandonment, because "I can't look at all the data from all patients" . In E2's view, however, monitoring technology is essential because "I need information, data which are certain, or a medium, or a platform with remote transmission". The fact that this technology can change "with artificial intelligence, fittingly" will allow:

> health to be monitored better, more detailed observations to be made, certain illnesses to be understood better, why they occur, in what context, etc.

In this context, i.e. if these technologies "are useful to me for monitoring patients" , then their use is entirely justified. A comparable point of view is found with E10, E11 and E12, for whom changes in technology must allow the patient to have more opportunities in terms of their pathology being treated (E10), or must allow more "medical time" for the doctor, encouraging better organization of the medical activity (E11). In E12's eyes, this change is irremediable:

> In addition, it will contribute more to us, as we said, with artificial intelligence even more so, and on the other hand connected objects for patient monitoring, in a general sense. [...] But at the same time, it's important to stay vigilant.

This requirement for vigilance comes hand in hand with the doctor's responsibility to their patients, a theme that respondents E6, E13 and E17 insist upon.

2.4.3. *The doctor's responsibility*

In E6's opinion, the doctor's responsibility is a mix of "intellectual rigor and pragmatism, always putting themselves in the position of an elderly and slightly confused patient [...] and with high quality treatment". It is not only a case of adapting the technology to the patient, but of changing all patient's medical treatments to avoid administrative red tape, on the one hand, and, on the other hand, to anchor contemporary medicine in its time:

> Either we are ready to adapt [...], or we remain stuck on our notion of medical practice in the 20th century, and nothing changes. In my opinion, that is dangerous.

This demand to "follow the movement" (E13) is accompanied by information to the patient, who must be kept informed and must provide their "free and informed consent". In E13's view, this involves being at the forefront of changes in medical technologies: "In the practice, we have always been very, very in-tune, connected, but that's because I take care of all that." This sense of responsibility towards patients motivates the doctor's investment in appropriation of technology: "the subject must be well-known in order for patients to be informed", because the doctor has obligations concerning their patients, which are imposed on them by the monitoring technology:

> He [the doctor] is not only responsible when he sets up surveillance, but it is also his duty to respond to alerts for example. [...] All these things must be set up.

This assertion of responsibility, indissociable from the desire to support patients, is both a requirement to avoid submitting to the technology ourselves or making patients submit to it. This last aspect was particularly noted by the respondents E4 and E18.

2.4.4. *The daily influence of technology*

E4 states: "I don't spend my days watching data flows, because it's impossible." In E4's eyes, technology is not able to replace the doctor–patient relationship, even though remote monitoring "is quite positive". On the contrary, E4 admits: "I believe we shouldn't look at it all the time..., it's not Big Brother, it's not a surveillance system." In E4's opinion, monitoring technology has its use and importance for the data it delivers, which allows us "to do our jobs as doctors". It must be used as a tool, nothing more. E18 argues along similar lines, noting the "time saved [...] on administrative tasks" allowed by the technology, while specifying, however: "I refuse to go and look at the data for each patient every day, once a month, or whatever." However, E18 specifies that, contrary to colleagues who are younger or more experienced than she is in information technology, she puts the relationship with the patient at the center of her activity:

> For them, anything that will be done to simplify their lives, and to avoid them from seeing sick patients, they will be happy with. For me, this is not the goal [...] I say to them "it's important to go and ask the sick person questions."

During our analysis, we observed that doctors did not share the same points of view on the three series of questions submitted to them, and they did not assign the same weight to them in their discussions. In the following sections, we examine this observation in greater detail. Our analysis by topic models enables us to understand the dimensions of the respondents' discussions, which are more or less supported than others. Regarding the content of the discussions, we established dual contributions – both from the doctors themselves and from their patients – in relation to the monitoring technology. What were the main findings from the point of view of the respondents?

2.5. Determining the legitimacy of monitoring technology

To understand what determines the legitimacy of monitoring technology from the respondents' point of view, we report on their positioning within the three series of approved questions, in order to obtain a visual representation of their overall position on all the topics analyzed so far. We also rename the topics into question categories, such that, for example, the first topic in the first series of question A becomes A_0, etc., for other topics and their corresponding question categories (see Figure 2.1).

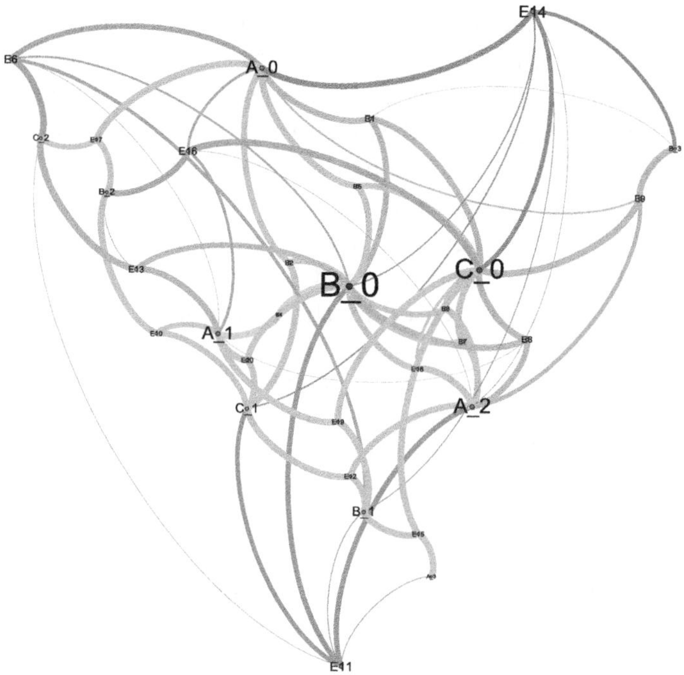

Figure 2.1. *Influence of the topics in the doctors' discussions*

We observe that the topics most supported by the respondents are B_0 and C_0, followed by A_2, then by A_1 and A_0. Referring back to the interpretation of these topics, we note that for the most part, doctors highlighted the theme of the doctor–patient relationship through the intermediary of monitoring technology and the data it generates. It is not surprising that the topic B_0 is the most supported by the respondents: it includes, in particular, other themes that we see in the second central theme, in other words, trust, development of the reticular universe based on this doctor–patient relationship covered by monitoring technology, and the relationship with service providers.

A specific aspect of topics A_2, A_1 and A_0 is that they are associated with the monitoring technology as it is perceived and used by the respondents. We have seen that these three topics are related to the themes of the effectiveness of monitoring technology, the verification it requires from doctors and the data it provides. For doctors, this technology and its use are connected to the challenge of successfully involving patients in the relationship with monitoring technology, which requires a collaboration between the doctor and the patient to verify the data and discuss the results delivered by the technology.

The monitoring technology serves as the formal medium through which a doctor and patient join together in a relationship mediated by technology, to make the treatment as effective as possible and improving the patient's health and well-being.

This challenge of willing participation is also a double challenge of legitimacy, namely the legitimacy of the technology in the eyes of both doctors and patients, which in turn guarantees the legitimacy of medical practice. As a consequence, for doctors, it is a case of serving more patients and potentially targeting other pathologies, while also developing their medical practice by integrating new expertise.

Is this general trend shared in the same way by all respondents involved in the study? The question is all the more relevant given that the respondents are public hospital personnel (E2, E4, E5, E6, E10, E12, E14, E18, E19, E20) and university hospital personnel (E1, E7, E9) in addition to self-employed doctors (E8, E11, E13, E15, E16, E17). Let us compare the way these groups of doctors support the topics linked to the three main themes of the questionnaire, starting with the first category labeled "A".

2.5.1. *Specialization versus multitasking*

Groups of doctors differentiate themselves from each other in the first question category, regarding their practices and their use of digital devices and connected objects, as indicated in the graphs below (see Figure 2.2).

We note that respondents working as self-employed medical personnel support all four topics, whereas professionals from public hospitals support only three of the four topics, while those in university hospitals support only two.

These results suggest that the context of medical practice affects the practice and use of monitoring technology by doctors, depending on whether they work in a self-employed environment or within a hospital structure. Respondents working as self-employed doctors are literally working on all fronts because their relative isolation requires them to remain up to date with all aspects of the monitoring technology. Public hospital professionals tend to leave the theme of the technology's development horizon to one side, which is not surprising given that they are part of a structure that directly influences the monitoring technology they will be able, or obligated, to use.

Similarly, doctors in university hospitals will focus above all on the technical and scientific aspects of the technology, with attention paid to the data it provides. At this stage, we have analyzed these profiles in the context of the second question category. It concerns the doctor–patient relationship in the context of the monitoring technology being used.

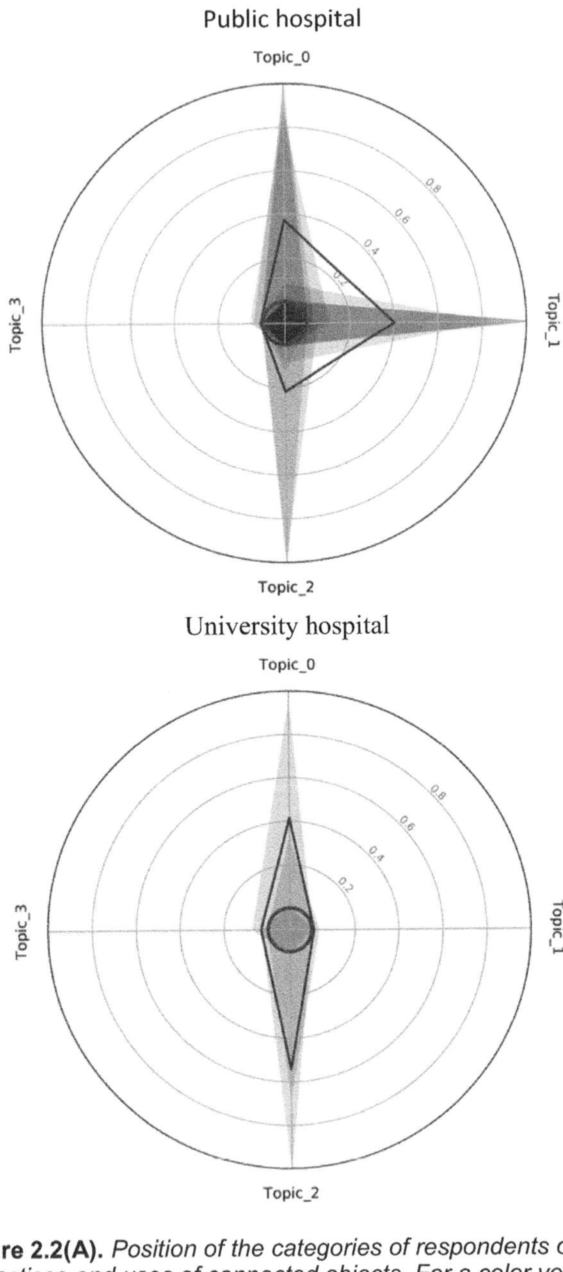

Figure 2.2(A). *Position of the categories of respondents on their practices and uses of connected objects. For a color version of this figure, see www.iste.co.uk/carre/digitaldevices.zip*

Figure 2.2(B). *Position of the categories of respondents on their practices and uses of connected objects. For a color version of this figure, see www.iste.co.uk/carre/digitaldevices.zip*

2.5.2. *Proximity of patients and delegation of the relationship*

As shown in Figure 2.3, the question of the doctor–patient relationship was strongly supported by all the respondents. Nevertheless, contrasts appear here depending on the environment in which they practice medicine.

The contrast is relatively weak between respondents in self-employed medical practices and those in public hospitals. Those working in public hospitals support all four topics. However, as shown by their average position indicated by the green line on the graphs, they tend to prioritize trust with the patient, by which it is assumed that an explanation has been given on how the monitoring technology works and what it can contribute in terms of improving their health condition. For self-employed medical professionals, this aspect is comparatively underdeveloped. On the contrary, they differentiate themselves from doctors in public hospitals insofar as they make use of the other apps which can be attached to the monitoring technology, and the way the patients tend to surround themselves with technologies to measure their health condition themselves, or at least collect on digital social media, the opinions of other patients who have pathologies similar to their own.

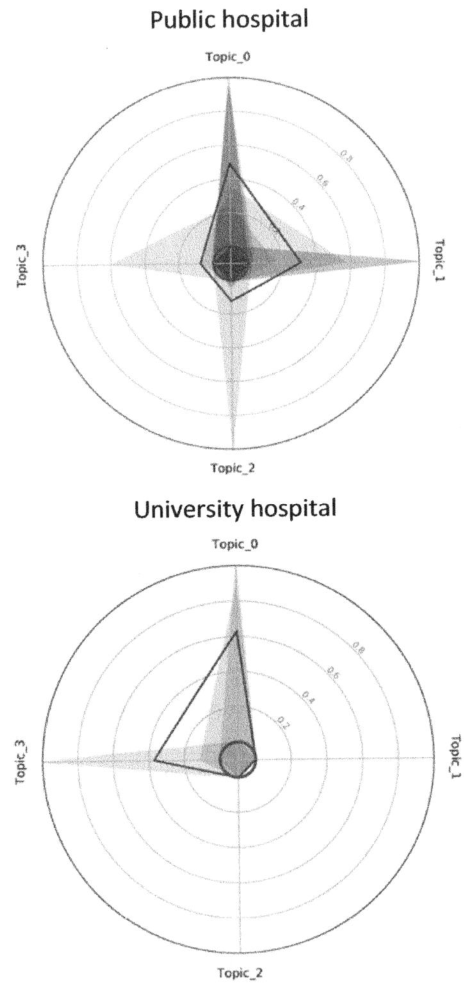

Figure 2.3(A). *Position of the categories of respondents on the doctor–patient relationship. For a color version of this figure, see www.iste.co.uk/carre/digitaldevices.zip*

Figure 2.3(B). *Position of the categories of respondents on the doctor–patient relationship. For a color version of this figure, see www.iste.co.uk/carre/digitaldevices.zip*

However, the most significant contrast emerges between these two categories of doctors and those working in university hospitals. While the former two categories of respondents develop the general topic of the doctor–patient relationship, the latter also supports the topic related to the service providers associated with the monitoring technology. These are, particularly, technicians from service providers who regularly visit patients to adjust the technology and ensure that it operates correctly. On this point, while public hospital professionals and self-employed doctors show they are closer to their patients, doctors in university hospitals tend to involve the service provider to a greater extent in the doctor–patient relationship, or to delegate to them part of the task of maintaining this relationship. Let us now examine the last question category related to the challenges of automation in the doctor–patient relationship.

2.5.3. *Responsibility*

The point of view held by doctors regarding the series of questions on the challenges of automation in the doctor–patient relationship also shows a less marked contrast between respondents who are self-employed practitioners and those in public hospitals, than between these two categories and doctors working in university hospitals (see Figure 2.4).

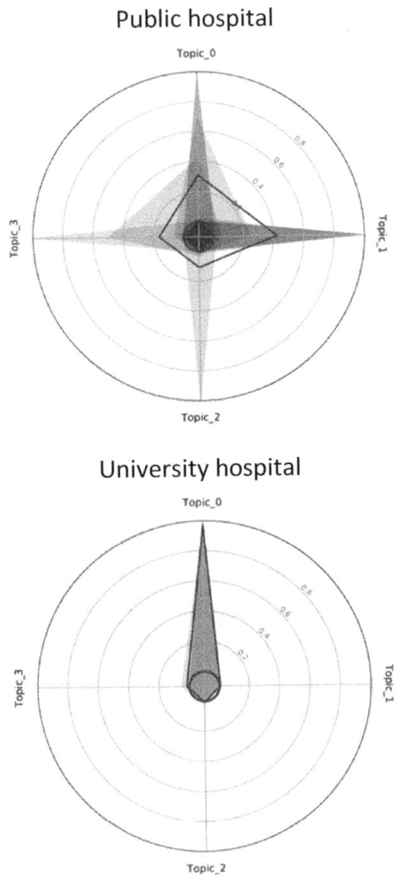

Figure 2.4(A). *Position of the categories of doctors on the challenges of automation in the doctor–patient relationship. For a color version of this figure, see www.iste.co.uk/carre/digitaldevices.zip*

Figure 2.4(B). *Position of the categories of doctors on the challenges of automation in the doctor–patient relationship. For a color version of this figure, see www.iste.co.uk/carre/digitaldevices.zip*

Doctors in university hospitals are the only ones to exclusively discuss the theme of the doctor–patient relationship in a context where treatment increasingly prioritizes technologies. Coupled with the two previous results, it is possible to deduce that for doctors in university hospitals, the connection between doctor and patient becomes even more important when mediatized by the monitoring technology. Concerning doctors of public hospitals, they differentiate themselves from the other respondents by occupying the four topics which structure this question category. On the one hand, they differ from self-employed doctors in terms of the attribution of responsibility, both their own responsibility as doctors and the responsibility of patients in remote monitoring. This theme is more dominant among respondents working as self-employed medical practitioners than among those in public hospitals. On the other hand, the latter differ from respondents who are self-employed practitioners as regards the theme relating to the importance of technology in their daily medical practice, a theme that is almost never discussed by self-employed practitioners. Once again, we can connect these contrasts between groups of doctors to the context in which they practice: public hospitals, on the one hand, with a comparatively greater throughput of patients and a large collegial and administrative structure, and, on the other hand, small medical practices with a smaller throughput of patients and a less preeminent collegial and administrative structure. These results must be repositioned within a broader context of the expansion of e-health in medicine.

2.6. Discussion

The medical field is traditionally partial to technologies, and the contributions assembled in this book are a reminder of this; see also (Wehling 2012; Huss 2019). It is not surprising to note that e-health is increasingly important in medicine (Andreassen and Skrøvseth 2016; World Health Organization 2021), as it is not only a question of associating technological progress and good health, which is essential in the medical field (Deutsche Plattform NanoBioMedizin 2020). In 2025, it concerns the data generated by these technologies; IDC Corporation estimated that in 2020, they represented 20% of all data generated around the world; see Gantz and Reinsel (2012), in other words, an indirect result, a by-product generated by their implementation. These technologies do not aim to treat the patient directly as medication would, but perform a continuous reading of their health condition. For a recent example in the field of sleep apnea, refer to Arnardottir et al. (2022); also see for more general considerations about self-quantification (Case 2017; Dagiral et al. 2019; Zouinar 2019). Such is the principal and more general message that our results convey: technology is no longer encapsulated in, or reduced to, a simple tool. It can generate scenarios, ideas and initiatives, and as such, it becomes a mediator that transforms the doctor–patient relationship by connecting the doctor to their patient.

Doctors perceive monitoring technology as a data-generating machine, at the heart of the doctor–patient relationship, which now inevitably extends to service providers and to other connected objects. Doctors, patients, monitoring technology (such as CPAP machines) and the connected objects that work with them guarantee the complete legitimacy of the monitoring technology and medical practice. This legitimacy is not granted; it is based on the willing use of the technology, which is at the center of relations between doctors, service providers and patients, as well as the willing participation of these actors in a connection to this technology and to the data it generates. This legitimacy creates expectations regarding the possibility of deploying the technology on a larger scale, in other words, to enable other patients to benefit from it or to apply it to other pathologies while maintaining the advantages it offers in terms of time savings and treatment precision. Even though the legal implications of using personal data to establish care strategies that are likely to improve the patient's health are not very clear today (Buchner and Hase 2016), for respondents to our study, it is the promise of being able to grow their medical practice, a promise of professional development, for example, towards medical research. In regard to patients, from the doctors' perspective, it is a promise to be able to benefit freely from their lives, without the restrictions associated with frequent medical visits or the repercussions of treatment that may not be suitable for changes in their pathology. Due to the data it generates, on the one hand, the monitoring technology formalizes the relationships between doctors, service providers and patients, but, on the other hand, it opens up spaces of emancipation for them, either directly through the attraction it creates, or indirectly, due to the problems it does not resolve, or even generate, which requires investment from the various actors in the relationship connecting them.

Concerning this theme of the doctor–patient relationship, doctors no longer view it without the monitoring technology. It is structured, on the one hand, by data generated by this technology, and, on the other hand, by the imaginary and the set of projections and strategies it liberates. There is a pedagogical aspect to it which rests with the doctor, in order to make the patient understand how the technology works, what its limits are, what the data collected directly by the CPAP machine means and that this machine has been able to enrich thanks to other connected objects or to the opinions gathered online and social media. The patient is more active and better informed about their health treatment and about their pathology. The relationship is modified, and for doctors, it presents a good opportunity to establish, based on this shared knowledge, a more fruitful working partnership with their patients. Nevertheless, from a different perspective, the doctor remains firmly attached to their role as an expert, which ensures that the patient receives treatment as optimal as possible, a guarantee that the monitoring technology can reinforce. Consequently, the doctors most involved in the monitoring technology are seeking to ensure they are surrounded by good service providers who are technically competent and comply with the protection of patients' data and the patients themselves. Willing participation by both doctors and their patients about the monitoring technology thus requires a significant degree of effort on the part of both doctors and patients, to ensure that the legitimacy they attribute to each other is maintained.

These challenges relating to the introduction of technologies in the doctor–patient relationship arise from the dual requirement for both investment in the technology and openness to technological developments that are to be integrated into medical practices in the name of responsibility that the respondents envisage as a dual guarantee of their medical practice. On the one hand, attention to technology offers, and in the future, will guarantee an improvement in the quality of patient care. On the other hand, since the doctor is invested in the technology and engages with their patients with it, they ensure that it does not obliterate their relationship with their patients, a relationship which remains a guarantee of the legitimacy of the relationship with the technology, both for the doctor and for the patient and, beyond this, of their legitimacy as practitioners-experts and patients.

While it appears that nothing can stop the spread of these technologies, with their indirect benefits – they generate data about patients' health, allowing care strategy profiles to be drawn up which place the pathology and changes in it that are traceable with patients being back at the center of considerations – they do not carry the same weight. The contrasts that we note regarding the positioning of the three categories of respondents on the themes mentioned (practices and uses of connected objects by doctors, relationship with patients in this context and challenges of connected objects for medical practices) during the interviews demonstrate this clearly. Doctors who practice in a public hospital context appear to be those for whom the challenge of monitoring technology, and beyond this, the technological challenge in general, seems to give rise to the most significant quantity of debates. Unlike the other two categories of doctors (self-employed and university-based), these doctors work in a context where the doctor–patient relationship is under tremendous pressure, particularly in terms of the number of patients to monitor. At the same time, the practice and use of technologies and connected objects do not depend on these doctors alone. They are also included in the context of developing the institution of a public hospital, which may be more demanding and even more restrictive than for self-employed practitioners or those working in university hospitals. Regarding self-employed doctors, the monitoring technology, which has become essential, is considered to be a necessary evil, with management delegated as much as possible to service providers. Conversely, doctors in university hospitals draw up scenarios which are likely to put this monitoring technology at the heart of a real relational device (Papilloud and Schultze 2022) by integrating other data-generating technologies capable of organizing them in the context of care strategy proposals (see the example of artificial intelligence mentioned by the doctors).

2.7. Conclusion

Our study focused on interviews with 20 doctors specializing in the treatment of sleep apnea, conducted within a global context of e-health development. These doctors now use monitoring technology, a connected object consisting of a CPAP device capable of reading and transmitting data about the patient's sleep. Our analysis using topic models identified three themes that structure all of the doctors' discussions: practices and uses of digital devices and connected objects (category A), relationships between doctors and their patients (category B) and the challenges of automation (category C). On these three themes, the doctors interviewed held different points of view, a contrast which arises from the differences in context and in medical practices. University hospital doctors are in favor of pairing the monitoring technology with other connected objects (their patients' watches or artificial intelligence in order to formulate diagnoses and treatment scenarios). Public hospital doctors have a more complex opinion. Indeed, this technology should not be rejected, but it should not replace the doctor–patient relationship. Lastly, self-employed doctors recognize the advantages of monitoring technology, but they only use it when truly necessary, on a case-by-case basis. These contrasts do not entirely mask a general trend of viewing the installation of this type of technology, of connected objects attached to it and potential extensions of it to artificial intelligence in a positive light. In the respondents' opinion, it is a change in medical practice which is seeking its own legitimacy. Doctors are now working in this expanding relational context, considering it both an opportunity to be seized and a challenge to rise to. It is a fabulous opportunity to gain a deeper understanding of the pathology affecting patients by analyzing data on an ever-increasing number of patients, to treat them more effectively. It is also a major challenge to maintain their status as health experts in the eyes of patients.

2.8. References

Andonova, Y., Vidal G., d'Ortho M.-P. (2023). Quand les objets connectés s'installent dans la pratique médicale. *Cordiam* [Online]. Available at: https://e-cordiam.fr/archives/20396 [Accessed 3 November 2024].

Andreassen, H.K. and Skrøvseth, S.O. (2016). Making sense of changing health landscapes. New demands for research and evaluations. *International Journal of Integrated Care*, 16(5), S1.

Arnardottir, E.S., Islind, A.S., Oskarsdottir, M., Olafsdottir, K.A., August, E., Jonasdottir, L., Hrubos-Strom, H., Saavedra, J.M., Grote, L., Hedner, J. et al. (2022). The Sleep Revolution project: The concept and objectives. *Journal of Sleep Research*, 31(4), 8. doi: 10.1111/jsr.13630 [Accessed 3 November 2024].

Buchner, B. and Hase, F. (2016). e-Health, Big Data und Co Informationelle Vernetzung der Medizin als Herausforderung für das Recht. *MedR*, 649.

Carré D., Vidal G., Sandré S. (2021). Recherche sur les objets connectés en santé (ROCS-SHS). Research Report, AP-HP, Paris.

Cases, A. (2017). L'e-santé : l'empowerment du patient connecté. *Journal de gestion et d'économie médicales*, 35, 137–158.

Dagiral, E., Licoppe, C., Martin, O., Pharabod, A. (2019). Le Quantified Self en question(s) : un état des lieux des travaux de sciences sociales consacrés à l'automesure des individus. *Réseaux*, 216(4), 17–54.

Deutsche Plattform NanoBioMedizin (2020). Aktionspapier 2020 NanoBioMedizin. Report, Deutsche Plattform NanoBioMedizin, Frankfurt am Main.

Gantz, J. and Reinsel, D. (2012). The Digital Universe in 2020 [Online]. Available at: https://www.cs.princeton.edu/courses/archive/spring13/cos598C/idc-the-digital-universe-in-2020.pdf [Accessed 3 November 2024].

Huss, R. (2019). *Künstliche Intelligenz, Robotik und Big Data in der Medizin*. VS-Verlag, Wiesbaden.

Lindenmeyer, C. (ed.) (2022). Enjeux et effets des dispositifs numériques sur la relation médicale – au prisme du traitement du syndrome des apnées du sommeil. Research Report, Institut La Personne en Médecine, Université Paris Cité and Centre Internet et Société [Online]. Available at: https ://cis.cnrs.fr/wp-content/uploads/2022/06/vf-rapport-recherche-cis-ilpem.pdf [Accessed 3 November 2024].

Papilloud, C. and Hinneburg, A. (2018). *Qualitative Textanalyse mit Topic-Modellen. Eine Einführung für Sozialwissenschaften*. VS-Verlag, Wiesbaden.

Papilloud, C. and Schultze, E.-M. (2022). *Skizze einer Theorie der Relation*. VS-Verlag, Wiesbaden.

Papilloud, C. and Schultze, E.-M. (2023). *Translation in der Nanomedizin*. Springer VS, Wiesbaden.

Wehling, P. (2012). Die Medizin auf dem Weg zur Technowissenschaft? Technowissenschaftliche Krankheitsidentitäten und die Schwierigkeit der Technikfolgenabschätzung. *Technikfolgenabschätzung. Theorie und Praxis*, 21(2), 15–21.

World Health Organization (2021). *Global Strategy on Digital Health 2020–2025* [Online]. Available at: https://iris.who.int/handle/10665/344249.

Zouinar, M. (2019). Théories et principes de conception des systèmes d'automesure numériques : de la quantification à la régulation distribuée de soi. *Réseaux*, 216(4), 83–117.

3

Digital Devices in Psychiatry and Mental Health

The advent of information technology in the 1940s did not fully cross over into the world of work until user accessibility improved. This came about when computer terminals became widespread in the 1970s, and later with the introduction of microcomputers and local networks in the 1980s. In the 1960s, companies began to computerize their administrative operations such as accounting, payroll and stock management, using applications that operated independently of each other (Volle 2015, p. 87). The Nora–Minc[1] report on the computerization of society emphasizes that telematics will potentially revolutionize the relational architecture of society as a whole. Indeed, unlike the technological innovations of the past (such as electricity or steam engines, for example), telematics conveys information, or in other words, power, according to Bergeron (1982, p. 155).

The digitalization of society from the 1990s onwards is very closely related to changes in electronic components, characterized by their constant miniaturization

Chapter written by Étienne HIEN.

1 This report on the computerization of society is published at the request of the President of the French Republic. Available at: https://www.vie-publique.fr/files/rapport/pdf/154000252.pdf [Accessed April 2, 2024].

(Jarrige 2016, pp. 289–310). According to the OECD[2], the term digital denotes a merging of products and services that capture, register and present data and information electronically. The digitalization of society significantly influences social life and its organization. Information and communication technologies (ICTs) are widely used and integrated into many sectors of society, thus encompassing the economy (banking, business, etc.), scientific research, logistics, culture, education (Ramel 2020, p. 18), as well as healthcare.

With the advent of datafication[3], digitalization has taken on a new dimension, transforming every aspect of daily life into a vast set of exploitable digital data (Mayer-Schönberger and Cukier 2013). This transformation, widely documented and analyzed in university research work (Boyd and Crawford 2012; Kitchin 2014), raises questions about privacy, data security, ethics and the socioeconomic implications of a digital society.

Initially based on the conversion of analog data into digital format, digitalization and computerization have laid the foundations for the "digital revolution" by enabling information to be stored, processed and transmitted rapidly and effectively (Castells 2000). Therefore, the computerization of the health sector manifests itself in the introduction of a large number of connected devices (blood pressure monitors, pedometers, etc.), health applications and websites, robots for the care of the elderly, projects that use artificial intelligence for diagnosis, information systems and remote care devices (Ramel 2020). This creates the concept of

2 The OECD regarding digitalization [Online]. Available at: https://www.oecd.org/content/dam/oecd/en/publications/reports/2025/06/oecd-definitions-of-the-ict-sector-the-content-and-media-sector-and-information-industries-based-on-isic-rev-5_455bfea7/b9576889-en.pdf?utm_source=chatgpt.com [Accessed December 16, 2024].

3 Datafication Wikipedia article [Online]. Available at: https://en.wikipedia.org/wiki/Datafication [Accessed December 16, 2024].

"digital health" (Mitchell 2000, pp. 16–19), which encompasses the combined use of the internet and information technologies for clinical, educational and administrative purposes, either remotely or locally. Stakeholders in the field (health professionals, public health policies and the WHO) view the integration of ICT in healthcare as a response to the challenges of health systems, an innovative solution, and a way to improve the performance of these systems. The increasing integration of connected digital devices in the field of health opens up new prospects, and has accelerated particularly since the Covid-19 pandemic, during which the use of effective digital tools became necessary according to the World Health Organization (WHO)[4]. However, this transition comes with consequences – data confidentiality, the reliability of the information provided by the applications and the impact on the patient–healthcare professional relationship raise concerns (monitoring, data use) (Torous and Keshavan 2016). As the WHO report emphasized, although these developments are significant, they carry risks, in particular concerning the human dimensions of care and fundamental rights. In addition, accessibility and acceptability of these technologies among various population groups within the wider population, including marginalized groups, require particularly close attention.

According to the ANS[5] (French Digital Health Agency), although digital health, or e-health, has recently become widely accessible through ICT, it is rooted in a history that goes back several decades, since the transformation of traditional medicine which occurred at the start of the

4 Empowerment through digital health [Online]. Available at: https://www.who.int/europe/initiatives/empowerment-through-digital-health [Accessed December 16, 2024].

5 Agence du Numérique en santé [Online]. Available at: https://esante.gouv.fr/les-50-ans-dhistoire-de-la-e-sante [Accessed December 16, 2024].

"computer revolution" of the 1970s[6], and became more evident with the emergence of the industrial digital health market in the 1990s. Access to medical information became increasingly widespread, as the power of the Internet grew in the 2000s. True awareness of the importance of digital health took place in the 2010s with the development of ambitious public policies, supported by a series of major investment plans, leading to an accelerated "democratization" of digital health in response to challenges posed by the health crisis in the 2020s. This sector is currently undergoing a major transformation due to the integration of IT and digital technology. Technological developments have paved the way for smarter healthcare, offering many advantages for both health professionals and patients.

In 2019, to face these new challenges experienced by the French healthcare system (Merlière 2016), in particular medical deserts, the aging population, the rise in chronic illnesses and polypathologies, and with the aim of meeting the needs of the population in terms of access to care, the French government has presented the roadmap "Accelerating the digital shift"[7], a project that was designed to put digital technology at the service of healthcare. In line with this commitment to innovation and modernization, in 2021, the "*Ségur Numérique*" program was launched to enable secure and seamless exchanges of health data between professionals and patients, with the goal of strengthening prevention and improving care. To do so, according to the digital health roadmap, a record funding of 2 billion euros has been allocated, including 1.4 billion over three years for sharing essential health data and 600 million over five years dedicated to the medical–social sector. These investments

6 Petite histoire de la e-santé [Online]. Available at: https://esante.gouv.fr/la-petite-histoire-de-la-e-sante [Accessed December 16, 2024].

7 See: https://sante.gouv.fr/IMG/pdf/feuille_de_route_du_numerique_en_sante_2023-2027_-_document_de_presentation.pdf [Accessed December 16, 2024].

are entirely funded by the European Recovery and Resilience plans[8].

3.1. Mental health and psychiatry

Considered for a long time as a taboo subject, mental health has become a political priority in recent decades, which has made it a public health issue. The World Health Organization regularly publishes data about the worldwide mental health situation. According to Béatrice Lamboy (2005, pp. 583–596), "the World Health Organization estimated that mental disorders affected approximately 450 million people worldwide". The Covid-19 pandemic highlighted the global mental health situation. In the WHO memorandum of June 17, 2022, it is written that in "2019, nearly a billion people – including 14% of adolescents around the world – were affected by a mental disorder. Suicides accounted for more than one in every hundred deaths and 58% occurred before the age of 50" (p. 583). Due to this pandemic, which caused several countries around the world to implement lockdowns for several weeks, innovative care initiatives (teleconsultation, telemedicine) were increasingly prioritized. However, we must note that these practices did exist before the Covid-19 pandemic, as indicated by Alexandre Mathieu-Fritz (2018, p. 123) – since the early 2000s there has been a significant increase in web-based offers of mental healthcare. These cover various practices of listening, advice, support, coaching and lastly "e-therapies", which can take the form of psychotherapy administered via email, instant messaging, teleconsultation (Convert and Demailly 2003; Demailly 2007) or even self-help devices that largely replace the face-to-face relationship with a psychotherapist (Bergström 2013).

8 Concerning the recovery plan, see: https://commission.europa.eu/strategy-and-policy/recovery-plan-europe_en [Accessed December 16, 2024].

The contributions of technology in the health sector have been broadly recognized, both by practical observations, scientific studies and the available empirical data (Akrich 2006; Klein and Borelle 2019). Digital tools play a significant role in the treatment of certain mental disorders: virtual reality has emerged as a promising tool for the treatment of phobias, post-traumatic stress disorders and other anxiety disorders (Freeman et al. 2017); activity trackers and mobile applications enable individuals to monitor their mental well-being on a daily basis by following indications such as sleep, physical activity and mood (Firth et al. 2017).

3.2. Projects and technical devices in psychiatry and mental health

Following the Covid-19 pandemic, the number of mental health apps increased exponentially, among which some stand out, such as Feel[9], which aims to associate psychology and technology. We will examine various projects and devices that have been implemented.

The PsyCARE university-hospital research project (RHU)[10] is funded by the future investment program (*Programme d'investissement d'avenir*), which aims to develop early intervention in psychiatry in France, particularly through prevention, detection of initial signs and personalized care. As part of this project, technological tools are deployed to collect data about patients, and sometimes non-patients, in order to better characterize the patient's

9 Feel, a self-therapy app for depression, designed with psychiatrists who are experts in cognitive-behavioral therapy (CBT). See: https://feelapp.care/ [Accessed April 2, 2024].

10 The project PsyCARE is one of the winners of the fourth wave of calls for projects from RHU. See: https://psy-care.fr/projet/investissements-davenir-et-projets-rhu/ [Accessed April 2, 2024].

pathology and develop customized treatment for them. Therefore, the *Scope-pilote* study (work package 6.1), conducted with 40 patients, evaluates the effectiveness of a cognitive training tool, its acceptability and user satisfaction. This study is performed in the context of a broader study, Scope (Self-administered COgnitive Personalized training in Early psychosis), a randomized controlled trial in adolescents and young adults to assess efficacy and efficiency of an eHealth application providing personalized cognitive training, funded by the French Ministry of Health's hospital clinical research program (PHRC).

New technologies offer opportunities to create user-friendly tools that avoid stigmatization, in particular apps that include cognitive strategies, motivation features and personalized advice tailored to the needs of every individual, according to those managing the PsyCARE university-hospital research program, directed by Professor Marie-Odile Krebs, head of a facility in the Paris university hospital group (GHU Paris Psychiatrie et Neurosciences), on the Sainte-Anne site, and coordinated by INSERM (French National Institute for Health and Medical Research). This project is supported by a national consortium coming from the research group of the French Psychiatry Institute, which brings together academic experts in neurosciences and bio-informatics, and clinical centers of excellence in the field of psychosis (Transition network), with the partnership of companies (Digital & Ethics and FEALINX). The advantage of the tools used is their accessibility to patients, wherever they are. The PsyCARE university-hospital research program is also developing a mobile app, working with the company ApyCare[11], whose objective is to provide a secure environment where patients can be supported through their mental healthcare journey. Through an instant messaging

11 ApyCare is a mobile app for health which develops a solution to improve communication and coordination of care by developing the patient's autonomy. See: https://apy.care [Accessed April 2, 2024].

service, they can contact the members of their care team and case managers in complete confidentiality; mental states and habits can also be monitored using self-assessment scales in the app, and treatment alerts can be received, for example, for medication or psychotherapy appointments.

Other technical devices are also being used in studies for the treatment of schizophrenia and psychoses, where virtual reality (VR) plays a key role. The study involves patients with schizophrenia participating in a virtual game, which takes place in a fictional town, over a three-month period. The Paris Psychiatry and Neurosciences University Hospital Group illustrates these advances in the field of mental health and psychiatry, implementing virtual reality as a form of therapy, in particular in addictology: "The patient finds themselves in a virtual universe face to face with the object of their addiction in an everyday context, such as an evening where cocaine is being passed around. The therapist can thus directly observe the patient's behavior in a craving situation and work with them to develop appropriate strategies," according to Raphaël Gaillard[12]. At this same hospital, the application of electrodes for the treatment-resistant OCDs (obsessive compulsive disorders) by deep brain stimulation (DBS) is undergoing a significant change thanks to the integration of new technologies. These advances are seen through the development of techniques based on brain/machine interfaces, which allow brain activity to be measured in order to regulate the relevant neural networks, at the right place and time.

Similar research using virtual reality (VR) has been conducted by researchers and clinicians, demonstrating its effectiveness in improving cognitive functions in

12 Raphaël Gaillard is a psychiatry professor and author of the book *L'homme augmenté*. See: https://www.ghu-paris.fr/fr/actualites/lia-en-psychiatrie-un-futur-qui-se-conjugue-au-present-au-ghu-paris [Accessed April 2, 2024].

schizophrenic patients. In a 2016 study, 10 patients affected by schizophrenia participated in a virtual game program, set in a fictional town, for three months. In a pilot study prior to this, seven schizophrenic patients attended weekly 90-minute sessions, during which they discussed the difficulties encountered in their daily lives. After familiarizing themselves with the virtual tools, such as a joystick and the virtual reality environment, they drew up strategies with which to overcome common obstacles, such as shopping or appointments, using alternative routes and switching from 2D to 3D maps (Amado et al. 2016).

Also noteworthy is the RevAlz study, which concentrates on the cognitive training of elderly people using tablet computers. The primary objective of this research was to design a serious game which would improve cognitive skills related to memory, in particular episodic and prospective memory, in healthy older people. The game offers a playful way to train these skills while supporting the transfer of gains to other cognitive areas, such as attention and executive functions. The study also evaluates the effectiveness of this tablet-based serious game in strengthening cognition in older adults. The hypothesis is that if the game is played regularly for a month, there should be significant improvements in episodic and prospective memories, and in other cognitive domains such as processing speed, executive functions and sustained attention. This study was carried out as a joint project by France Alzheimer, the Paris University Hospital Group, the French Institute for Memory and Alzheimer's Disease and the *grande école* Normale Supérieure (ENS) (Piolino 2022).

The study "Determination of EEG microstates associated with mental disorders in at-risk states" (DEMETER) also illustrates these technological innovations in mental health and in psychiatry within the Paris University Hospital Group. The study collects data from individuals with mental disorders (including mood disorders, psychoses and autism spectrum disorders) as well as healthy participants. Data from 126 subjects (aged between 15 and 30 years) on speech, sleep and sense of self have been collected and analyzed, using information technology software. The participants underwent extensive phenotyping based on psychopathology, neuropsychological assessments, EEG recordings, biological sampling, sensorimotor tasks, speech recording, polysomnography, and performed a virtual reality task (Lucarini et al. 2023). The study is ongoing. We also note that with the increasing power of new technologies and deep learning through algorithms and AI, it is now possible to predict depression. This is the result of work performed by a team of researchers at the University of Caen (Milintsevich et al. 2023).

For example, algorithms – defined as "a series of instructions leading to a result" (Hallé 2016, p. 74) – are logical structures underlying computational processes applied to massive databases, known as big data. They organize and select content according to users' online information practices, as shown in Dominique Cardon's work (2015). Algorithms, such as Google's PageRank and Facebook's NewsFeed, are used to organize and prioritize content on online platforms, including search engines and social media networks. Online activities and the use of connected devices generate vast amounts of information about individual behaviors, consumption habits and political or religious preferences. The term "big data" encompasses all these data, whether they originate from the Internet, social media networks, corporations or connected devices, and represent a key challenge in the analysis and use of information (Menger and Paye 2017). The work of Michael Birnbaum et al. (2016) on the impact of online resources and social media on help-seeking behaviors among young people presenting psychotic symptoms illustrates effectively how, via online research and social media, information about these users is collected and analyzed.

These therapies, facilitated by connected and digital devices, involve using information and health data with the aim of developing personalized care. According to the French Data Protection Agency (*Commission nationale de l'informatique et des libertés* – CNIL), the French Health Data Platform (*Plateforme des données de santé* – PDS) and the European Union's Health Data Hub, there is an aim to facilitate the sharing of health data from various sources in order to encourage research. The CNIL defines personal health data as information about the physical and mental health of a person (past, present or future), including their medical history, examination results and treatments. By cross-referencing with other data, this collected information can allow conclusions to be drawn about a person's health status[13]. The circulation of health data, unless pseudonymized and anonymized, can compromise privacy and jeopardize the confidentiality of sensitive information. In 2020, the French State Council, aware of the risks of transferring large quantities of data on European citizens to the United States, requested guarantees from the Health Data Hub to limit these risks. Indeed, "since Microsoft was given the task of hosting the platform, a complaint was filed by a range of associations and professionals to the French State Council to request suspension of the Health Data Hub, due to the recent ruling of the European Union Court of Justice (CJEU) of 16th July 2020"[14]. The platform, which aims to facilitate the sharing of health data from different sources in order to encourage research, received opinions and recommendations from the CNIL, presented on January 31,

13 See: https://www.cnil.fr/fr/quest-ce-ce-quune-donnee-de-sante [Accessed April 2, 2024].

14 See: https://www.cnil.fr/fr/le-conseil-detat-demande-au-health-data-hub-des-garanties-supplementaires [Accessed April 2, 2024].

2024[15]. The regulatory authority is required to authorize the use of health data[16].

To engage in the discussion on technological determinism and solutionism, it is essential to highlight how digital innovation in health raises complex questions about the decisive role of technologies and risks.

3.3. Innovations in health: the question of technical determinism

The concept of technical determinism emerged in the 1970s in correlation with the development of information and communication technologies (ICTs), thus placing technology at the center of societal concerns. According to Doray and Millerand (2015a, p. 66): "For several years, the sociology of technologies and the sociology of work examined the status of tools and objects in the analysis of technological development, and their impact on social life." This concept provoked heated controversies within humanities and social sciences (HSS), which have examined the relationship between innovation and the technical offer. The authors (Alsène 1990, pp. 321–337) highlight several points of view associated with the controversies: one view suggests that technical progress affects life at work. Organization, for example, is believed to be shaped not only by the techniques used, but also by the strategies of the actors, the methods of managing technological change and the type of economic market in which the company operates. This concept of determinism implies that the organization of work is influenced not only by the technologies used, but also by the strategies of the actors involved, the methods of managing technological

15 See: https://www.cnil.fr/fr/les-principaux-avis-et-recommandations-de-la-cnil-sur-la-plateforme-des-donnees-de-sante [Accessed April 2, 2024].
16 See: https://www.health-data-hub.fr/ [Accessed April 2, 2024].

change and the economic context (Doray and Millerand 2015b, pp. 66–69).

Sociologists have also questioned whether technical determinism has been the driving force behind history. The deployment of these tools in society has changed practices in several aspects of life, including in the world of work. In the health sector, in particular mental health, these tools constitute significant possibilities for improving patient care. Since the 1960s, medical informatics specialists have published research supporting the idea that introducing information technologies can improve healthcare quality[17]. However, technical determinism is questioned by researchers – according to Proulx, limiting the impact of a technology to a single factor (technical or social) would be reductive, because political or social change results from a series of mediations between human actors and technical devices (Proulx 2001, p. 58); similarly, Josiane Jouët highlights a "double mediation of the technical and the social" which establishes a connection between social and technical innovation, used as an analysis framework for uses of ICTs (Jouët 2000, p. 497). Subscribing to technical determinism implies perceiving technology as the "savior of humanity", an innovation that compensates for human shortcomings when facing difficulties. With this in mind, some, including the historian Hughes (1983) and the computer scientist Kling (1991), have focused their research on demonstrating that society also influences techniques, as emphasized in the work of Melvin De Fleur (1966, pp. 314–326): "While most studies of mass communication attempt to unravel ways in which the media influence society, the present analysis tries to bring out ways in which society has influenced the media." Proulx et al.

17 Silber, D. (2004). *The Case for eHealth*. European Institute of Public Administration [Online]. Available at: https://www.researchgate.net/profile/Denise-Silber/publication/8013927_The_case_for_eHealth/links/5593db7008aed7453d477092/The-case-for-eHealth.pdf

highlight the existence of a reciprocal influence between social aspects and digital tools. The social web, for example, carried forward by internet developments from 2004 onwards, is characterized by functionalities that encourage users to work together to create and share content (Millerand et al. 2010, p. 2). The authors mention developments in web-based architectures and applications which encourage users to work together to create and share content, including tools such as blogs, wikis, social media networks (Facebook, LinkedIn, etc.), sharing sites (MySpace, Flickr, YouTube), and metaverses (Second Life). Indeed, the social web creates a space for exchanges and sharing in which users actively participate, thus underscoring an interdependence between technologies and social aspects. In his book *The Whale and the Reactor: A Search for Limits in an Age of High Technology*, in 1986 Langdon Winner (Rivière 1987) tackles technical determinism – he explores social and political implications of modern technologies, criticizing the tendency to consider technological solutions as universal remedies for social problems, while neglecting the more complex and nuanced aspects of social issues (Rivière 1987, p. 131). In the health sector, we note that there is a certain form of dehumanization of medicine with the loss of the human dimension associated with the rise of these technological tools:

> The practice of evidence based medicine means integrating individual clinical expertise with the best available external clinical evidence from systematic research. By individual clinical expertise we mean the proficiency and judgment that individual clinicians acquire through clinical experience and clinical practice (Sackett et al. 1996, pp. 71–72).

3.4. Innovations in health: between technical solutionism and social control?

Considering the multiple dimensions of digital tools, it is difficult, according to Steve Moradel[18], the author and specialist in spatial computing and AI, to deny the significant benefits they have brought to our lives and the potential for improvements in living conditions in many regions of the world. However, in his view, the omnipresence of technology and its impact on billions of individuals' lives, in addition to the interconnectedness of everything at all times, raise some important questions. These include questions of protection of privacy and personal data, and equal access to technologies and information. These questions highlight the complex challenges and opportunities associated with the ubiquity of technologies in everyday life. Although presumed to enhance accessibility, it is essential to note that it can also reinforce inequalities. Indeed, some populations have difficulty accessing these technologies and mastering them, which creates a social divide (Régnier 2012), whereas this progress was supposed to provide solutions for all.

In his book *Pour tout résoudre, cliquez ici*[19] published in 2014, Morozov provides a definition of the concept of technological solutionism. The author identifies two ideologies associated with the development of new technologies: on the one hand, solutionism, promoted by the actors of Silicon Valley, who believe that technology can improve all aspects of life and resolve most problems; on the other hand, webcentrism, a new obsession in digital economics, which places digital technology at the center of the world. This drive towards digital technology has led to

18 Chronicle on the limits on technical solutionism, in 2022; see: https://www.stevemoradel.com/post/les-limites-du-techno-solutionnisme [Accessed April 2, 2024].

19 Morozov, E. (2014). *Pour tout résoudre, cliquez ici: l'aberration du solutionnisme technologique* . Fyp éditions, Limoges.

several negative consequences such as algorithmic monitoring, the illegal use of user data, targeted advertising abuses and identity theft.

Today, governments and institutions tend to adopt an approach based on technological solutionism in their quest to solve problems, without necessarily understanding their underlying causes. A striking example of this trend is the Stop Covid-19 application, strongly criticized for its potential use for the purposes of "digital surveillance"[20] of the population, particularly regarding people's movement. These tools, according to Vidal (2022), are used to exercise control and influence the behavior of individuals and objects, thus justifying the need for surveillance. Information technologies allow vast quantities of data to be processed, often provided by the users themselves, as underscored by the famous expression "unbeknownst to them". Digital tools have unquestionably introduced a series of developments, as well as provoked an array of concerns regarding the protection of privacy. Among the major challenges are the practices of collection and use of data by large technological companies – according to Christl and Spiekermann (2016), some of these companies can gather up to 30,000 data points for each individual that they monitor.

In the health sector, algorithms and artificial intelligence, particularly machine learning, offer significant advantages by processing data and by providing valuable information to healthcare professionals. This improvement positively impacts patient outcomes and well-being. Among some digital tools supporting mental health diagnosis is the PHQ-9, a questionnaire developed in 1999 by Drs. Spitzer, Williams and Kroenke. Comprising nine questions, it enables

20 Olivier Tesquet: "Je crains plus l'installation d'une surveillance généralisée que Stopcovid" [Online]. Available at: https://www.rue89strasbourg.com/olivier-tesquet-surveillance-generalisee-stopcovid-175699 [Accessed April 2, 2024].

screening and assessment of depressing severity. While widely used in clinical settings and sometimes available online, this test does not replace the expertise of a healthcare professional for a formal medical diagnosis (Kroenke and Spitzer 2002).

More precisely, structuring of patients' data facilitates diagnosis, guides therapeutic care and contributes to the prediction of suicidal behaviors and neurodegenerative diseases. Despite these advances, concerns remain: security measures and AI remain vulnerable to attacks, with even minor handling operations which may lead to erroneous results (Finlayson et al. 2019, pp. 1287–1289). Beyond the advantages of these technologies for health, they are also sources of inequalities, and the lack of regulation (despite the AI Act) remains a cause for concern in our societies:

> What becomes clear is that across diverse domains and contexts, AI is widening inequality, placing information and control in the hands of those who already have power and further disempowering those who don't.[21]

This report from 2019 points out the commercial incentives and conflicts of interest which have made ethical and legal questions regarding health data into news items: "Most recently, a whistle-blower report alerted the public to serious privacy risks stemming from a partnership, known as Project Nightingale, between Google and Ascension, one of the largest nonprofit health systems in the US."

Technology as a solution to societal problems has not yet been sufficiently detailed, because many cases of use of these technologies invite us to rethink their use, as demonstrated

21 AI Now (2019). AI Now 2019 Report [Online]. Available at: https://ainowinstitute.org/wp-content/uploads/2023/04/AI_Now_2019_Report.pdf [Accessed April 2, 2024].

by this unfortunate situation in the United Kingdom: a 14-year-old girl committed suicide after being exposed to a large amount of content on suicide and depression on websites such as Instagram. Some of this content was algorithmically selected for her, based on the platform's tracking of her viewing habits[22]. The use of AI and digital technologies to monitor workers, in particular by controlling the length of their breaks, may encourage them to avoid these necessary breaks and neglect social interactions. This creates a competitive and stressful environment, leading to anxiety and stress among workers (Rapport du Mental Health Europe 2022, p. 10)[23].

The work of Foucault on monitoring devices (*dispositif*) is relevant here to extend our discussion. The emblematic panopticon model, presented in *Discipline and Punish* (1975), is presented as a power mechanism, combining architecture and monitoring. According to the author, panoptics maintains individuals in a constant state of visibility, thus ensuring power can operate automatically (Gonzalez 2015, pp. 11–17). At the Collège de France, in his class entitled "Security, territory, population" (1977–1978), Foucault explored the various methods of the devices, in particular the legal, disciplinary and security mechanisms. He underlines their interconnection and the emergence of statistical practices to define normality and abnormality. Additionally, although little mention of it was made by Foucault at the time, the development of information technology devices can also be viewed as electronic monitoring devices exercising administrative control, and thus included in the rationale of power devices.

22 See: https://www.judiciary.uk/wp-content/uploads/2022/10/Molly-Russell-Prevention-of-future-deaths-report-2022-0315_Published.pdf [Accessed April 2, 2024].

23 Mental Health Europe (2022). La santé mentale à l'ère numérique. Report [Online]. Available at: https://www.mhe-sme.org/wp-content/uploads/2023/04/La-sante-mentale-a-lere-numerique.pdf [Accessed April 2, 2024].

3.5. Economic challenges and health data security

Health data are subject to strict regulations, such as the General Data Protection Regulation (GDPR), to guarantee their protection. However, despite these measures, since 2018 the CNIL has authorized[24] many pharmacists to collect pseudonymized personal data from millions of French citizens. These data are then transmitted to IQVIA, a major American data broker, as reported by Steve Moradel on his website[25]. He specifies that although IQVIA claims not to use these data for commercial purposes, the risks remain. Technology companies and insurers invest massively in this sector, as shown by the partnership between the British National Health System and Alphabet for processing the data of 1.6 million patients. This raises concerns about security, integrity and confidentiality of the collection and processing of patients' data. A notable example is the Grindr[26] app, which was involved in a scandal for having shared the personal data of its subscribers, including their HIV status, with companies. We also note that recent authorization of the use of health data on the Microsoft platform (Health Data Hub) by the CNIL[27] raises concerns among French people[28], as with this authorization, data from

24 See: https://www.cnil.fr/sites/cnil/files/2023-09/guide-rgpd-cnop_cnil.pdf [Accessed April 2, 2024].

25 See: https://www.stevemoradel.com/post/les-limites-du-techno-solutionn isme [Accessed April 2, 2024].

26 Le temps (2018). L'appli Grindr accusée d'avoir partagé des données privées [Online]. Available at: https://www.letemps.ch/cyber/lappli-grindr-accusee-davoir-partage-donnees-privees [Accessed April 2, 2024].

27 CNIL (2024). Plateforme des données de santé: missions et enjeux [Online]. Available at: https://www.cnil.fr/fr/plateforme-des-donnees-de-sante-missions-et-enjeux [Accessed April 2, 2024].

28 France Info (2020). Données de santé des Français: le choix contesté de Microsoft [Online]. Available at: https://www.francetvinfo.fr/internet/securite-sur-internet/donnees-de-sante-des-francais-le-choix-conteste-de-microsoft_4126165.html [Accessed April 2, 2024].

hospital centers, pharmacies, the "My health account" online space, and research data will be stored at Microsoft Azure, Microsoft's cloud-based platform. Some French companies see in this a loss of economic opportunity and also a threat or a loss of digital sovereignty, while others, such as Adrien Parrot[29], a doctor and computer scientist, see neglect of digital freedom following this decision by the CNIL.

3.6. Conclusion

Despite the influence of technical determinism and technological solutionism in healthcare, it is essential to highlight that doctors should retain the final decision-making role. Although progress has been made in artificial intelligence and machines can provide valuable information and decision-making tools, they must not replace the clinical decisions made by human medical expertise, including in mental health and psychiatry. Decisions must consider multiple factors, not only medical data but also individual and subjective aspects unique to each patient:

> In the context of an individual-based approach, emphasizing listening and empathy, the use of chatbots, from the earliest ones such as Eliza to the most sophisticated developed by Silicon Valley startups, already calls established practices into question (Aimé 2017, pp. 51–56).

According to Aimé, for the moment, these devices do not aim to replace practitioners. However, these devices clearly perform their role effectively, and individuals, whether sick or healthy, often interact with them as if speaking to a human being. Ultimately, the relationship between doctor and patient, based on trust, empathy and medical expertise,

29 See: https://www.dailymotion.com/video/x80tchn [Accessed April 2, 2024].

should remain essential. Continuing Aimé's analysis (2017), it is useful to state that in psychiatry this positioning is all the more essential given that recent developments in psychiatry, combining biological and heterogeneous data for personalized medicine guided by data (Data Driven Medicine), are developing an integrative approach. Whereas the flow of medical data is ever increasing, human limits in big data management become apparent, thus emphasizing the potential of AI technologies in processing and analyzing this data rapidly and precisely. Although psychiatry is still in its infancy, other medical specialties such as oncology and ophthalmology have already used this advanced approach to adapt and improve treatments (Aimé 2017, p. 56).

Following Bernard Stiegler's analyses, it would be possible to describe digital technologies in healthcare, whether for mental health or other areas, as pharmakon: "We believe that digital technology, specifically, is pharmakon – a poison and a cure" (Stiegler 2015, p. 16). Indeed, moving towards a digital health economy significantly changes the traditional paradigms of care services, creating new economic models and new opportunities for intervention. On this subject, Stiegler (2015, p. 18) states:

> The digital industry is a data economy, and it operates by tracking and thereby capturing the activity of internet users through extraordinarily highly-developed tracking systems.

However, this transformation is not without risk, particularly regarding the protection of sensitive personal data and health information security. Additionally, technical determinism and technological solutionism can lead to dehumanization of healthcare, by relegating the role of the doctor to second place behind algorithms and machines.

3.7. References

Aimé, X. (2017). Intelligence artificielle et psychiatrie : noces d'or entre Eliza et Parry. *L'information psychiatrique*, 93, 51–56. doi: 10.1684/ipe.2017.1582.

Akrich, M. (2006). Les objets techniques et leurs utilisateurs. De la conception à l'action. In *Sociologie de la traduction*, Akrich, M., Callon, M., Latour, B. (eds). Presses des Mines, Paris. doi: 10.4000/books.pressesmines.1198.

Alsène, E. (1990). Les impacts de la technologie sur l'organisation. *Sociologie du travail*, 32(3), 321–337. doi: 10.3406/sotra.1990.2513.

Amado, I., Brénugat-Herné, L., Orriols, E., Desombre, C., Dos Santos, M., Prost, Z., Krebs, M.O., Piolino, P. (2016). A serious game to improve cognitive functions in schizophrenia: A pilot study. *Frontiers in Psychiatry*, 7, 64. doi: 10.3389/fpsyt.2016.00064.

Batifoulier, P. and Da Silva, N. (2021). L'économie encastrée dans la santé. Introduction au dossier. *Revue française de socio-économie*, 26, 13–26. doi: 10.3917/rfse.026.0013.

Bergeron, J. (1982). L'informatisation de la société : une réflexion critique autour de quelques ouvrages récents. *Politique*, 1(2), 153–165. doi: 10.7202/040407ar.

Bergström, J. (2013). Les nouvelles thérapies par Internet. *RESET*, 2 [Online]. Available at: http://journals.openedition.org/reset/124 [Accessed 15 December 2024].

Birnbaum, M.L., Candan, K., Libby, I., Pascucci, O., Kane, J. (2016). Impact of online resources and social media on help-seeking behaviour in youth with psychotic symptoms. *Early Intervention in Psychiatry*, 10(5), 397–403. doi: 10.1111/eip.12179.

Boyd, D. and Crawford, K. (2012). Critical questions for big data: Provocations for a cultural, technological, and scholarly phenomenon. *Information, Communication & Society*, 15(5), 662–679. doi: 10.1080/1369118X.2012.678878.

Brisset, N. (2014). Performer par le dispositif ? Un retour critique sur la théorie de la performativité. *L'Année sociologique*, 64, 217–246. doi: 10.3917/anso.141.0215.

Capital (2021). Pourquoi le dossier médical partagé peine à séduire en France [Online]. Available at: https://www.capital.fr/economie-politique/pourquoi-le-dossier-medical-partage-peine-a-seduire-en-france-1410690 [Accessed 2 April 2024].

Cardon, D. (2015). *À quoi rêvent les algorithmes. Nos vies à l'heure des big data*. Le Seuil, Paris. doi: 10.3917/ls.cardo.2015.01.

Castells, M. (2000). Toward a sociology of the network society. *Contemporary Sociology*, 29(5), 693–699. doi: 10.2307/2655234.

Chambaretaud, S. and Hartmann, L. (2004). Économie de la santé : avancées théoriques et opérationnelles. *Revue de l'OFCE*, 91, 235–268. doi: 10.3917/reof.091.0235.

Christl, W. and Spiekermann, S. (2016). *Networks of Control: A Report on Corporate Surveillance, Digital Tracking, Big Data & Privacy*. Facultas, Vienna.

Convert, B. and Demailly, L. (2003). Internet et les professions de santé : le problème de la consultation à distance. *Réseaux*, 120(4), 241–269 [Online]. Available at: https://shs.cairn.info/revue-reseaux1-2003-4-page-241?lang=fr.

De Fleur, M.L. (1966). Mass communication and social change. *Social Forces*, 44(3), 314–326. doi: 10.2307/2575832.

Delpierre, C. (2018). Les nouvelles pratiques liées au big data en santé : enjeux sociétaux et impact au regard des inégalités sociales de santé. In *Santé, numérique et droit-s*, Poirot-Mazères, I. (ed.). Presses de l'Université Toulouse Capitole, Toulouse. doi: 10.4000/books.putc.4434.

Demailly, L. (2007). Que fait l'Internet aux psychanalystes ? *Le Carnet PSY*, 120(7), 30–34. doi: 10.3917/lcp.120.0030.

Doctrine du numérique en santé (2022). [Online]. Available at: https://esante.gouv.fr/sites/default/files/media_entity/documents/doctrine-du-numerique-en-sante_version-2022_vf.pdf [Accessed 2 April 2024].

Doray, P. and Millerand, F. (2015a). Déterminisme technologique. In *Sciences, technologies et sociétés de A à Z*, Bouchard, F., Doray, P., Prud'homme, J. (eds). Presses de l'Université de Montréal, Montreal, Quebec. doi: 10.4000/books.pum.4280.

Finlayson, S.G., Bowers, J.D., Ito, J., Zittrain, J.L., Beam, A.L., Kohane, I.S. (2019). Adversarial attacks on medical machine learning. *Science*, 363(6433), 1287–1289. doi: 10.1126/science.aaw4399.

Firth, J., Torous, J., Nicholas, J., Carney, R., Rosenbaum, S., Sarris, J. (2017). The efficacy of smartphone-based mental health interventions for depressive symptoms: A meta-analysis of randomized controlled trials. *World Psychiatry*, 16(3), 287–298. doi: 10.1002/wps.20472.

Foucault, M. (1975). *Surveiller et punir*. Gallimard, Paris.

Foucault, M. (2004). *Sécurité, territoire, population. Cours au Collège de France (1977–1978)*. Gallimard/Le Seuil, Paris.

Freeman, D., Reeve, S., Robinson, A., Ehlers, A., Clark, D., Spanlang, B., Slater, M. (2017). Virtual reality in the assessment, understanding, and treatment of mental health disorders. *Psychological Medicine*, 47(14), 2393–2400. doi: 10.1017/S003329171700040X.

Gonzalez, A. (2015). Le dispositif : pour une introduction. *Marges*, 20, 11–17. doi: 10.4000/marges.973.

Greenan, N. and Mairesse, J. (2006). Les changements organisationnels, l'informatisation des entreprises et le travail des salariés : un exercice de mesure à partir de données couplées entreprises/salariés. *Revue économique*, 57, 1137–1175. doi: 10.3917/reco.576.1137.

Hallé, P. (2016). À quoi rêvent les algorithmes ? *Annales des Mines – Gérer et comprendre*, 123, 74–76. doi: 10.3917/geco1.123.0074.

Hughes, T.P. (1983). *Networks of Power: Electrification in Western Society, 1880–1930*. Johns Hopkins University Press, Baltimore.

Jarrige, F. (2016). *Technocritiques. Du refus des machines à la contestation des technosciences*. La Découverte, Paris. doi: 10.3917/dec.jarri.2016.01.

Jouët, J. (2000). Retour critique sur la sociologie des usages. *Réseaux*, 100(2), 487–521 [Online]. Available at: https://shs.cairn.info/revue-reseaux1-2000-2-page-487?lang=fr.

Kitchin, R. (2014). Big data, new epistemologies and paradigm shifts. *Big Data & Society*, 1(1). doi: 10.1177/2053951714528481.

Klein, N. and Borelle, C. (2019). Réalité virtuelle et santé mentale. *Revue d'anthropologie des connaissances*, 13(2), 1–4 [Online]. Available at: http://journals.openedition.org/rac/1275.

Kling, R. (1991). Computerization and social transformations. *Science, Technology and Human Values*, 16(3), 342–367.

Kroenke, K. and Spitzer, R. (2002). The PHQ-9: A new depression diagnostic and severity measure. *Psychiatric Annals*, 32, 509–521. doi: 10.3928/0048-5713-20020901-06.

Lamboy, B. (2005). La santé mentale : état des lieux et problématique. *Santé Publique*, 17(4), 583–596. doi: 10.3917/spub.054.0583.

Lucarini, V. et al. (2023). Neurophysiological explorations across the spectrum of psychosis, autism, and depression, during wakefulness and sleep: Protocol of a prospective case-control transdiagnostic multimodal study (DEMETER). *BMC Psychiatry*, 23(1), 860. doi: 10.1186/s12888-023-05347-x.

Mathieu-Fritz, A. (2018). Les téléconsultations en santé mentale : ou comment établir la relation psychothérapeutique à distance. *Réseaux*, 207, 123–164. doi: 10.3917/res.207.0123.

Mayer-Schönberger, V. and Cukier, K. (2013). *Big Data: A Revolution that will Transform How We Live, Work, and Think*. Houghton Mifflin Harcourt, Boston [Online]. Available at: https://psycnet.apa.org/record/2013-17650-000.

Menger, P.-M. and Paye, S. (eds) (2017). *Big data et traçabilité numérique. Les sciences sociales face à la quantification massive des individus*. Collège de France, Paris. doi: 10.4000/books.cdf.4987.

Mental Health Europe (2022). La santé mentale à l'ère numérique. Report [Online]. Available at: https://www.mhe-sme.org/wp-content/uploads/2023/04/La-sante-mentale-a-lere-numerique.pdf [Accessed 2 April 2024].

Merlière, Y. (2016). Le renouveau du dossier médical : Big Data et objets connectés en santé. *La Revue des Sciences Sociales*, 50(2), 84–90. doi: 10.3906/aps-sant-2016-2-1.

Milintsevich, K., Sirts, K., Dias, G. (2023). Towards automatic text-based estimation of depression through symptom prediction. *Brain Informatics*, 10(1), 1–14.

Millerand, F., Proulx. S. Rueff, J. (eds) (2010). *Web social. Mutation de la communication.* Presses de l'Université du Québec, Quebec.

Mitchell, J. (2000). Increasing the cost-effectiveness of telemedicine by embracing e-health. *Journal of Telemedicine and Telecare*, 6(1), S16–S19. doi: 10.1258/1357633001934500.

Morozov, E. (2014). *Pour tout résoudre, cliquez ici. L'aberration du solutionnisme technologique*. Fyp éditions, Limoges.

Piolino, P. (2022). La réalité virtuelle, pour une neuropsychologie de la vie quotidienne. In *Pratiquer les cyberpsychothérapies. Jeux vidéo. Réalité virtuelle. Robots*, Tisseron, S. and Tordo, F. (eds). Dunod, Paris, 143–150.

Proulx, S. (2001). Usages des technologies d'information et de communication : reconsidérer le champ d'étude ? In *Actes du 12e Congrès national des sciences de l'information et de la communication* (SFSIC).

Ramel, V. (2020). Les technologies numériques en santé face aux inégalités sociales et territoriales : une sociologie de l'action publique comparée. PhD Thesis [Online]. Available at: https://theses.hal.science/tel-02963193/document.

Régnier, F. (2012). Technologie numérique et impact socioculturel. *Hegel*, 1(1), 23–37. doi: 10.3917/heg.021.0023.

Rivière, J. (1987). Langdon winner. The whale and the reactor, a search for limits in an age of high technology. *Revue française d'études américaines*, 31, 131 [Online]. Available at: www.persee.fr/doc/rfea_0397-7870_1987_num_31_1_1895_t1_0131_0000_2.

Sackett, D.L., Rosenberg, W.M., Gray, J.A., Haynes, R.B., Richardson, W.S. (1996). Evidence based medicine: What it is and what it isn't. *BMJ (Clinical Research Ed.)*, 312(7023), 71–72. doi: 10.1136/bmj.312.7023.71.

Scala, B. (2016). E-santé - La médecine à l'ère du numérique. *Science&Santé*, 29, 22–33.

Silber, D. (2004). The case for eHealth. European Institute of Public Administration [Online]. Available at: https://www.researchgate.net/profile/Denise-Silber/publication/8013927_The_case_for_eHealth/links/5593db7008aed7453d477092/The-case-for-eHealth.pdf.

Stiegler, B. (2015). Demain, le temps des automates et le temps de la désautomatisation. *Intellectica*, 63, 151–160. doi: 10.3406/intel.2015.1028.

Torous, J. and Keshavan, M. (2016). The role of social media in schizophrenia: Evaluating risks, benefits, and potential. *World Psychiatry*, 15(3), 220–227. doi: 10.1002/wps.20303.

Vidal, G. (2022). Informatique et société. Research Report, LabSIC-USPN.

Volle, M. (2015). Comprendre l'informatisation. *Cahiers philosophiques*, 141(2), 87–103. doi: 10.3917/caph.141.0087.

4

Connected Devices in Health: Between Mobilization, Experimentation, Assessment and Data Protection

Connected devices in the healthcare field, and more broadly digital technologies as a whole, constitute an organizational disruption. Deployment of these technical tools requires expertise, new professions and decision-making paths, in particular within health establishments. These devices are positioned within the highly regulated field of healthcare.

Covid-19 highlighted the use of connected objects in healthcare for both monitoring patients and protecting users' personal data, not forgetting the personalized monitoring of each user during this worrisome time:

> The increase in the number of health apps has been exponential: 35,000 health and well-being apps (from fitness to meditation), with record numbers of downloads (2.5 billion) during the

---

Chapter written by Sarah SANDRÉ.

Covid-19 pandemic and a generation, those in their thirties, who are particular fans of these health apps (France Mobiles 2022; Parizeau 2023).

Hospitals and public institutions encourage digital solutions to be rolled out through the intermediary of many incentive-based mechanisms: funding, calls for projects, calls for expressions of interest, creation of third places (incubators, acceleration programs), partnerships and collaborations with doctors or hospital departments. However, a gap can be observed between "techno-scientific promises"[1], the aura that connected objects enjoy, the true efficiency of certain connected devices and the difficulty of implementing them within a department (legal risk management, ease of learning to use them, patient information, etc.). Despite the concerns expressed by some on digital technologies since the Covid pandemic regarding protection of individual freedoms or even questions about the fate of the personal data collected, digital technology remains a strategic theme in the field of health. This priority is all the more evident in a context where the notion of a "lack" is increasingly characterized in health: a lack of appeal and personnel in the hospital, medical deserts and drug shortages. A hospital is also the theater of many experimentations including an increase in third places and partnerships with start-ups.

We are going to examine in more detail the implementation methods for potentially innovative connected devices in hospital settings. To this end, we will rely on the roadmaps "France 2030" and "My Health 2022" which describe the strategy for accelerating policies for the development of digital health in France. Other than setting up platforms supporting public/private partnerships (G-NIUS), there are

1 We have borrowed and translated this expression from Pierre-Benoît Joly.

various sources of funding for health innovation. These roadmaps contribute to the implementation of experimentation with connected devices (in particular in terms of remote monitoring). A deafening observation can be added to this; the humanities and social sciences are entirely absent (or perhaps have just a marginal presence) in the establishment of a doctrine and specifications for innovation in digital health. The term "innovation" is used here generically in reference to the aforementioned institutional documents which use this notion.

4.1. The appeal of connected objects in the field of health

Technical innovation evokes progress. These two notions bear the promise of an improvement[2]. This prospect is particularly strong in the health sector, where needs are increasingly important. In the eyes of the government, digital technology appears to be the solution to numerous problems such as the monitoring of chronically ill patients, an aging population or even medical deserts. We indicate that the necessary use of digital technology in healthcare is relayed by institutional actors in this sector (health establishments in particular), as well as by companies marketing connected devices.

2 Here, we have restated the reasoning of the philosopher Thierry Menissier: "Viewed in a general manner, innovation constitutes an intellectual category employed to mean a change that aims to make an improvement. When viewed in a precise sense, it denotes the economic dynamics expressed in liberal societies through the production of objects and the design of new services available commercially and which create value; this acceptance is itself determined by technological invention correlated to the acceptance by consumers of objects and services produced thanks to the continuous emergence of new tools, procedures and methods" (Menissier 2016, p. 47).

4.1.1. *An incentive-based policy concerning the deployment of connected objects in healthcare*

On October 18, 2021, the French government released its "acceleration strategy" for the deployment of digital health. Taken from the "France 2030" roadmap, this strategy makes 650 million euros available to support the development, validation and experimentation of digital tools for 5P-medicine[3]. This budget allocates in particular 81 million euros to training all actors in the digital health sector and 20 million euros per year to setting up calls for projects assessing the medical benefit and economic impact of digital or artificial intelligence-based medical devices on new digital uses in health. It should be noted that 65 million euros are set aside for the creation of third places (incubators, experimentation laboratories such as fab labs, etc.) and 50 million euros for reinforcement of innovation-assistance devices targeting new uses of digital technologies in healthcare. We note that public investment in healthcare has increased considerably since the pandemic. The roadmap "Innovation Santé 2030" coordinated by the French Strategic Council of Health Industries (CSIS) has developed a public investment doctrine which is applied to all health actors including hospitals. Financial support of 7 billion euros dedicated to health innovation is thus envisioned. In this case innovation is understood generically in regard to biomedicine, clinical trials, support for industry, company growth in the health sector and, of course, the digital technology sector (650 million euros).

The French government possesses an arsenal of schemes used for rolling out digital health with a strategy based on:

A broad public consultation (429 answers) and targeted interviews (46 interviews) which have

3 So-called 5P-medicine: personalized, preventive, predictive, participatory and precision.

> enabled the ecosystem to express its expectations and its priorities. These answers have specified and prioritized the content of strategic acceleration, and participate in the design of future calls for projects (French government and CSIC 2021).

This momentum has for several years been at the origin of a significant increase in calls for projects and public–private partnerships which aim to encourage the development of digital health. In addition to these measures, it should be noted the desire of all healthcare actors to have clarity on the supply of connected health objects in France. The influx of connected objects on the French health market requires a structuring of the supply, both for political actors, potential financial backers of these digital tools, and it means user confidence can be ensured (patients, health professionals, assistants). The Ministry of Health, using the roadmap "*Ma Santé 2022*", set up the French national bureau of innovation and uses in e-health, also known as "G-NIUS". It supports project innovators wishing to bring digital tools to the market in the healthcare sector. This is a declarative process in which project innovators connect with this device. G-NIUS thus makes information available, designed to be used by entrepreneurs. The site includes several statutory documents relating to compliance of digital devices designed for the health sector; in particular, if the tools will be medical devices, and of course statutory compliance regarding the protection of health data, including of course the GDPR. All these resources illustrate a logical process whereby actors self-regulate. This bureau also draws up a list of all available funding. It is a database listing all possible funding that can be requested depending on the degree of maturity and type of project concerned. The geographical area (company headquarters, for example) influences the available funding in particular in areas affected by medical deserts[4]. These multiple examples strengthen the promise of progress based on digital technology. We could even refer to the "aura" of digital technology. To quote the philosopher Marcello Vitali-Rosati in his *Éloge du Bug*, "What is immaterial is better and belongs to another world" (Vitali-Rosati 2024, p. 107). In addition to being a tool providing assistance for the development of digital health projects, G-NIUS aims to be a hub, otherwise known as a platform grouping together all digital health actors in France. Structuring and legibility are thus an integral part of the deployment of digital health devices in France. However, these imperatives do not provide any clarity of the definition of innovation. Indeed, how can specifications be drawn up to define what can be considered innovative in the digital health field? Very concretely, an institution needs clear lines, guidelines and decision-making paths to determine which device it is best to use.

The G-NIUS platform describes several funding mechanisms that project backers can call on, including reimbursement schemes[5]. Among these mechanisms, there is the experimentation device enshrined in article 51 of the French Social Security Financing Act (LFSS). This article establishes the deployment of innovative solutions, generally remote monitoring devices. It also encourages experimentation with these solutions, in particular in hospitals. Start-ups and hospitals establish a partnership on the basis of this article which is paid for by French national health insurance. Implementation of this solution does not cost the hospital anything. For the project backer, they must comply with the

4 G-NIUS, the French national bureau for innovation and uses in e-health, "trouvez une aide financière pour votre projet e-santé". See: https://gnius.aides-territoires.beta.gouv.fr/ [Accessed February 24, 2024].

5 "Since it was launched in 2020, G-NIUS is the service that facilitates the path forward for health innovators. It saves time for all concerned by answering your questions, by directing you towards the right contacts, and by helping you identify sources of funding." French government (2024). France 2030, two years of the "digital health" strategy: making France a leader in digital health. Press release, Thursday January 18, 2024, 20.

specifications imposed by this article and provide a kind of "proof of concept"[6]. One purpose of article 51 of the LFSS is to assess with the hospital (according to the agreement drawn up between the project backer and the establishment) and with the Regional Health Agency (ARS):

> The objective of article 51 is to promote innovative organizations that contribute to improving patient treatment pathways, the efficiency of the health system, access to care, or even the pertinence of prescriptions for health products. Whilst health pathways are progressively being put in place to ensure comprehensive patient care, this is a real opportunity for new approaches to be tested. This device allows us to break away from many common law funding rules, applicable both in cities and in hospital or medical-social establishments[7].

Assessment and experimentation are an integral part of the digital health acceleration strategy in France because they are necessary for structuring the supply of connected objects in the healthcare sector. Experimentations and assessments are also conducive to an increase in partnerships between digital technology actors and health actors, as illustrated by the increase in calls for projects in light of the government roadmap "France 2030": 445 project backers spread across France, and 148 winning projects between 2022 and 2024[8]. However, it is not easy to establish assessment grids or experimentation plans, in particular in hospitals. This complexity stems from the difficulty in defining and

6 An experimentation by which the feasibility of a project or of rolling out a device can be confirmed. This is also referred to as POC.

7 See: https://www.iledefrance.ars.sante.fr/article-51-innovation-organisationnelle [Accessed 24th February 2024].

8 Gouvernement (2024). France 2030, 2 ans de la stratégie "santé numérique" : faire de la France un leader de la santé numérique. Press Release, January 18, 2024, 9.

clearly qualifying what innovation in digital health actually is, and in ensuring several expert opinions interact: health professionals, engineers, information technology security managers, project managers specialized in the digital transformation, etc.

4.1.2. *"Intermediate" actors, relays of this doctrine?*

Calls for projects generate partnerships between traditional actors in healthcare, namely public health establishments (nursing homes, public or private hospitals), research structures (Institut Pasteur, INSERM, CNRS, research laboratories with specialization in health, universities) and public institutions (regional health agency, health insurance offices), public authorities which work to reduce medical deserts (health centers, etc.) and digital technology actors from the private sector (pharmaceutical laboratories which develop digital tools, or even the private sector, for example, INRIA). The creation of an environment suitable for connected objects in health is illustrated by the strategy of developing "third places" by the public sector. The "France 2030" roadmap planned for a call for projects "Third places for experimentation" to begin on February 23, 2022:

> Assessment and experimentation constitute a major theme for this strategy in order to encourage the emergence of innovative solutions, supported by multidisciplinary scientific approaches and ambitious medical-economic models, to conquer the digital health market while it is developing worldwide. Third places for experimentation bring together a broad range of actors (health professionals, users, patients and/or healthy volunteers, assistants, authorities, companies, suppliers of digital solutions, researchers, methodologists, economists, those in

> charge of promotion, etc.) to implement experimentations, evaluate them and contribute to their access to the market (Secrétariat général pour l'investissement 2022).

This call for projects involves a three-phase project selection process for projects managed by the *Banque des territoires* on behalf of the French government. The objective is to create places where digital health solutions can be experimented with and assessed: testing the impact on users, evaluating the effectiveness of connected objects, performing studies of user needs, measuring the medical and economic benefits of the solutions tested, supporting deployment and market access for solutions having demonstrated their impact and associating professionals and those concerned in co-designing solutions (Secrétariat général pour l'Investissement 2022). Third places for experimentation act as laboratories serving digital health. The appeal of these spaces is particularly high for health establishments. Establishing a third place for experimentation is considered to have many advantages: testing a connected object under real conditions, connecting digital experts and doctors, identifying a hotbed of innovations, legitimizing startups by hosting them in a third place at the hospital, raising the visibility of companies hosted for potential investors (fundraising) and providing doctors with the opportunity to test a digital solution within their department. In 2021, the Assistance Public des hôpitaux de Paris (AP-HP) launched a project for a third place for experimentation, @hoteldieu, comprising an incubator (a startup incubator), the Digital Medical Hub (DMH), a platform to evaluate digital health tools created by the AP-HP in addition to a third place related to AP-HP[9] research structures. The intention is that third

9 AP-HP (2022a). Future4care, @Hôtel-Dieu et PariSanté Campus annoncent un partenariat inédit au service de la santé numérique. Press Release, March 9, 2022 [Online]. Available at: https://www.aphp.fr/

places for experimentation will be set up throughout France. The *Hospices civils de Lyon* (Lyon Civil Hospices – HCL) announced on January 18, 2024 that their platform PLATINES (*PLATeforme INnovation Environnement Santé*) received an award. This platform joined the first 15 third places selected in round 1 of the call for "third places for experimentation" projects in digital health run by the *Banque des territoires* in the context of the digital health acceleration strategy and of the initiative France 2030, and has thus been awarded 63 million euros[10]. In the same way as @hoteldieu or DMH, PLATINES aims to be a multi-disciplinary platform for evaluating and supporting digital solutions in the health field:

> After initial evaluation of the project by a multi-disciplinary team, the project is developed and optimized so that an evaluation suitable for the project requirements can be carried out (clinical testing, medical and economic assessment, environmental impact assessment). Finally, a co-development method is put in place, allowing the relevant key actors to collectively design the project with a view to optimizing it (HCL 2024b).

It is still too early to evaluate the effectiveness of this policy which has a clear aim: making France a European leader in digital health. The increase in calls for projects and consultations expresses the need for structuring and evaluation on a protean offer of connected objects, to which is added the uncertainty associated with artificial intelligence

---

contenu/future4care-hotel-dieu-et-parisante-campus-annoncent-un-partenariat-inedit-au-service-de-la [Accessed February 25, 2024].

10 HCL (2024a). La plateforme PLATINES labellisée "tiers-lieu d'expérimentation" en santé numérique . Press Release, February 1, 2024. [Online]. Available at: https://teamhcl.chu-lyon.fr/la-plateforme-platines-labellisee-tiers-lieu-dexperimentation-en-sante-numerique [Accessed February 25, 2024].

and with uses that patients, health professionals and assistants make or will make of the solutions developed and supported by the various devices set up by “France 2030”.

The government’s structuring strategy includes public consultations for certain connected medical devices. This involves the public consultation “digital medical devices and aging well”, for which a consultation has begun, and “digital medical devices in mental health” declared in October 2023. This consultation took place from January 18 to March 31, 2024 to help define a framework for digital innovations in the field of autonomy and aging[11]. Another consultation is dedicated to medical devices in the field of mental health, “digital medical devices in mental health” of the innovative medical devices[12] plan.

The rise in the number of these initiatives plays a significant role in the development of connected devices in the health field, in particular in hospitals, where the development of digital devices becomes both an imperative and a solution to simplify and improve the treatment of patients by health professionals and administrative workers. But do we really need digital technologies this much? Are digital technologies really a solution to everything?

Looking again at the notion of Mozorov’s “technological solutionism”, Marcello Vitali-Rosati (2024) is also questioning the systematic use of highly developed technologies. Indeed, do these connected devices really meet

11 Gouvernement (2024). France 2030, 2 ans de la stratégie “santé numérique” : faire de la France un leader de la santé numérique. Press file, January 18, 2024, 7.

12 “The objective is to bring about and maintain a context suitable for digital medical devices in order to ensure the prevention of mental illnesses, the identification and treatment of patients suffering from these disorders”, *ibid.*, p. 8.

the needs of their users, in particular in such a sensitive field as health?[13]

4.1.3. *Promises of connected objects for monitoring chronic illnesses*

During the 2010s, connected objects became widespread, to the point of becoming an essential tool for patients suffering from chronic illnesses, including caregivers. Popularization of the Internet of Things (IoT) was due to several factors. Firstly, as described above, the viral nature of this IoT was encouraged by public authorities: an increase in calls for projects, implementation of various funding devices, encouragement of partnerships and collaborations between actors in digital technologies and in health in particular via "third places for experimentation". The second element relates to the uses associated with connected objects in health. Chronic illnesses imply, on the one hand, long-term patient monitoring, and, on the other hand, patient cooperation in their own treatment. Digital technologies make a connection between health professionals, patients and caregivers. Most connected objects in the health sector are based on several promises: personalized care, patient empowerment, patient cooperation in their own care and understanding of their illness. Diabetes, sleep disorders, obesity or even monitoring of transplant patients are a good illustration of these characteristics:

> Expectations are very high: empowerment, better access to care, support for the shift to outpatient care for patients affected by chronic illness, monitoring of the safety of elderly patients or those with reduced autonomy, monitoring of the secondary effects of new medication in phase IV,

13 "In other words: we let ourselves be convinced that we have needs that are more and more costly in terms of technology, needs that only advanced technologies can meet" (Vitali-Rosati 2024, p. 173).

etc. ((Lindermeyer and d'Ortho 2019) quoted in (Parizeau 2023)).

One of the specific aspects of connected objects in healthcare is the possibility for the patient to measure their own data, for example, to control and analyze their glycemic curve using an insulin pump connected to an application. The quantified self has permanently changed practices in health and well-being, allowing health data to be collected in real time and the hyper-personalization of care to be possible. One consequence of the quantified self is self-tracking, i.e. the monitoring of our own data:

> Self-tracking devices and the "quantified self" may then also appear to arise from a form of "self-managerialization", through which people internalize managing obligations as "soft" prescriptions (in the sense that they do not appear to be directly imposed by external authorities) (Dagiral et al. 2019).

Self-tracking is a typical feature of connected health objects : the promise of improved patient monitoring is conditioned by the patient responsibility and, to a certain extent, monitoring. However, the act of measuring the data and their state of health requires the patient to be a loyal user of the connected object. The patient's willing involvement in their care pathway, capturing their attention requires working on the ergonomics and architecture of the connected solution. The user experience is an important part of an application's development. The architecture of an application can influence the user's choices, in particular in the way they collect their data. The "nudge" is a cognitive bias that can influence user behavior and choices regarding their digital solution. Linda Cambon (2016), a researcher in public health, specifies that "a "nudged" person is potentially pushed in a direction that contradicts their general system of values or preferences (if determined) and therefore, what they are made of. There is therefore nothing to say that this behavior will continue. It is more likely that they will re-integrate their original system of preferences, leading to a risk of disorientation, compensation and fragmentation of the personality (divided, contradictory, illogical person). These features of connected devices in digital health reveal a “functional imperative” which is an integral part of the “technical and scientific promises” borne by these objects. This imperative is illustrated in app functions: sending data in real time, improving the quality of a diagnosis, accelerating and simplifying the patient’s treatment conditions. In this paradigm, innovation in digital health “is what ensures everything works better” (Vitali Rosati 2024)[14].

4.2. Complexity of rolling out connected objects

There are general definitions for connected objects in health, and more broadly digital health and connected health devices. The growth in digital solutions in the health market, in addition to constant changes in uses brought about by users and by France’s ambition of becoming a leader in this field, have led to a shift in the definition of connected health devices. Confidence in the digital health market is determined by the effectiveness of the connected devices that are available.

Evaluation is not only a guarantee of the legitimacy of digital health, but also a guarantee of effectiveness for the patient. An evaluation also constitutes management of potential legal risks presented by new generations of digital solutions, including devices using artificial intelligence.

14 “A technology works when it accelerates a process; it does not work when it is not sufficient – the hypothesis of slowing something down is not even envisaged” (Vitali-Rosati 2024, p. 23).

4.2.1. *Evaluating connected objects in health: a precursor to legal risk management*

The DMH at the AP-HP has developed expertise in evaluating digital health solutions. As indicated above, connected devices target chronic illnesses or well-being issues. Sleep disorders are at the crossroads between care (chronic illness, sleep apnea, other pathologies related to sleep disorders) and wellbeing (stress management, sleep quality, weight management associated with sleep). The DMH carried out the study Evalexplo in order to evaluate the effectiveness of several connected devices on the market that are accessible to the wider public, in other words not medical devices only available on medical prescription. The main investigator, doctor Justine Frija Masson, presented her results for devices treating sleep disorders in November 2023 to the Sleep convention, co-organized in Lille by the French Society for Medicine and Research on Sleep (SFRMS) and the French Language Society for Pneumology (SPLF). They show disparities between connected objects sold to the wider public and those prescribed by doctors in a clinical context. The study thus compared "wider public" devices, in other words connected devices which are not "medical devices" according to the dispositions of French Public Health laws. Evalexplo thus studied the reliability of several connected weighing scales, blood pressure monitors, various activity trackers (a pedometer or a connected bracelet, for example), a thermometer and even a saturometer[15]. Doctor Justine Frija Masson evaluated two very popular devices, namely the Withings and Fitbit connected watches. The study was carried out with 61 patients at the AP-HP; 55.7% of

15 "In the proposed research, we are going to study the quality and reliability of the measurements provided by various connected objects sold to the wider public. To do so, we wish to compare the results provided by the device, with the results obtained during your examination, using the reference measurement methods ("gold-standard")." See: https://www.aphp.fr/registre-des-essais-cliniques/evaluation-de-la-fiabilite-metrologique-dobjets-connectes-disponibles [Accessed March 3, 2024].

them were men, with an average age of 54 years, whose examinations showed suspected sleep apnea (54%) or hypersomnia (26%) and who mainly complained of insomnia (26%) (Do Quang 2023):

> This Withings watch does not detect stages 1 and 2 very well, qualifying it as "light sleep" in addition to stage 3, and does not detect REM sleep, so of course this will affect the results. And conversely, the Fitbit overestimates the duration of light sleep and above all the duration of deep sleep, which will be very problematic. With the Fitbit, people will not understand why they are tired whilst in fact this deep sleep is not sufficient, and conversely, with the Withings, others who come for appointments believing they are not getting enough deep sleep. Words reported from Dr Justine Frija Masson, in 2023 (Do Quang 2023).

If patients suffering from a chronic illness (and if relevant, their caregiver) are going to follow their doctor's prescriptions, they will be attentive to any devices that allow them to live with their pathology more easily and improve their quality of life, including those intended for the wider public. The lack of precision of certain smart watches such as those mentioned in the study participate in the "gadget" effect that these objects can have and which holds back the legitimization process in digital health. Dr. Frija Masson reminds us that these devices cannot and must not be used in a healthcare context[16]. However, from the user's point of view and without a critical perspective, confusion between

16 "Globally, this independent study, conducted 'quite a large number of patients, more than in manufacturers' metrological studies', suggests that connected devices for the wider public are not very reliable in people with sleep disorder symptoms with respect to the polysomnography, even if it was only carried out on two subjects, in a single night, with patients consulting in a medical center, commented the researcher," *ibidem*.

care and wellbeing is increasingly recurrent. The technical evaluation of digital health tools requires several different areas of expertise to overlap. Research projects analyzing connected devices that use self-tracking group together several disciplines, in particular those from the humanities and social sciences (sociology, anthropology, information and communication sciences, etc.).

The Som'Health project has evaluated the Apple Watch and the Fitbit watch in the context of sleep disorders, just like the DMH at the AP-HP[17]. Emotional design, common to both the Apple Watch and the Fitbit watch, is pointed out by the researchers. This design is characterized, for example, by vibration of the bracelet or structuring of the alerts which condition the interactions between users and the device[18]. The use of gamification and nudges to make using this object fun is common and normalizes activity tracking and the anxiety it can generate: performance anxiety or orthorexia as it can be represented on social media networks (clean girl routine, video content on lifestyle hygiene in nutrition, for example, with the videos "what I eat in a day"):

> This sensory method of interaction based on vibrations is also seen in the Fitbit bracelet, which invites the user to rise to health challenges

17 The project is in particular directed by several researchers with a specialization in information and communication sciences, in particular Pierre-Michel Riccio, an engineer at the CNAM with a PhD in the subject. The project also unites designers and neurologists and sleep doctors. See: https://projetsomhealth.wordpress.com/equipe/.

18 "The first represents a promise of interaction and assistance through the intermediary of connected objects, by relying on emotional design, and publicity objects that embody a technical and central vision of the interaction between connected objects, users and their environment. The second is used massively amongst the population, in particular for sleep monitoring, and proposes a relationship based on play, thanks to a design approach based on the modification of user behavior" (Catoir Brisson et al. 2022, pp. 161–192).

> in a fun way. But the sleep monitoring functions are associated with other types of recorded data (for example, heart rate), which call into question the support for data reading outside of all medical monitoring [...] Performance anxiety can be caused by the comparison with a sleep norm depending on age and sex entered by the user at the start of the experiment. Additionally, the overabundance of notifications of the Fitbit bracelet can be intrusive in daily life (Catoir Brisson et al. 2022).

Evalexplo or Som'Health show us that we are still very far from being able to separate the wheat from the chaff in terms of connected objects designed for the wider public, but clearly target monitoring and an ambition for improving the health of its users. The lack of clear regulation for these objects poses a problem for users and potentially for assistants. The humanities and social sciences would have their full place in the evaluation devices, for example, to measure the risk of cognitive saturation of a device or to put some distance between the promise made about the device and its true effectiveness.

4.2.2. The imperative of protection of health data – law as an evaluation tool

With the exception of cyberattacks, the public is only rarely informed of the value of its personal data. However, the use and collection of health data are inherent to IoT (the Internet of Things). In this case, health data can be defined broadly: it is not only data about the patient's medical condition such as information about their treatment, their glycemic curve or their temperature, but it can also be data providing information about their living conditions, mental wellbeing or the user's personal life. Health data enjoy protection reinforced by the General Data Protection

Regulation (GDPR) which came into effect in May 2018 and by the provisions of French laws on public health regarding in particular the provisions on the confidentiality of medical information, on the conservation time thereof, or even, for example, on medical confidentiality[19]. While these provisions appear far-off for the patient, the French National Commission for Data Protection (CNIL) or even the French Agency for Digital Health (ASIP santé) draw up reference frameworks and guidelines to support the compliance of actors in digital health. Guaranteeing the security of patient data is a major challenge in a context where cyberattacks particularly target health establishments and organizations which hold health data (insurance organizations, for example). The choice of healthcare data host companies illustrates GDPR imperatives with regard to data processing managers, which are integrity, confidentiality, availability and security:

> A reinforced cloud data management strategy must therefore be set up, [and] it will take on the role of insurance policy for health organizations. If a public hospital is in the process of being restructured, it must also coordinate itself more with actors in grass-roots medicine via a platform of digital services which may be effective only by relying on the expertise of the Information and Telecommunications Technology (ITT) actor (Quémémer et al. 2020, p. 65).

Whilst data host companies must have a "health data host" authorization before processing health data in France, not all theses companies are created equal. "Companies" belonging to the GAFAM[20] such as Microsoft Azure or AWS

19 Here, the idea is not to be exhaustive but to demonstrate the importance of health data protection and the turning point represented by the Covid pandemic.

20 As a reminder, GAFAM refers to the multinational companies that dominate the digital technologies sector: Google, Apple, Facebook, Amazon and Microsoft.

(Amazon Web Service) are inspiring increasing suspicion among IT managers or data protection officers within public health establishments. This situation may generate tension during negotiations with startups who wish to establish a partnership with a hospital: indeed, how can a "startupper" be encouraged to change health data host company in the expectation of a partnership or an experimentation within a hospital? Digital technologies constitute an organizational overhaul for health establishments. This is an almost forced acceleration where data protection or even contracts that frame the experimentation and impose measurement of the impact of digital solutions (the French Health Agency calls for projects, for example, or experimentation in a hospital department) are the last acceptable safeguards to regulate digital technology. Due to its slow pace, the law appears to be less functional than digital technology: it is even considered to be an obstacle. Concern with regard to health data host companies is increasing following the Cloud Act, the American law enacted on March 23, 2018 on the use and conservation of health data. This law results in extraterritorial effects: in other words, an American access provider whose server is based on a European Union country can transfer data to the government at the request of the latter. Many startups are in the habit of using providers which belong to GAFAM. Hospitals adapt their GDPR compliance with adapted clauses (sovereign cloud) and several American companies in digital technologies increase the initiatives to reassure French institutions in particular in the health sector:

> AWS engaged in a partnership with the French company Atos in the cloud, without indicating if this alliance has the same objective. "France represents one of our main markets in Europe. Approximately 80% of companies in the CAC 40 use the AWS services", specified the managing director of AWS France, recalling that his group

> intends to invest 5.3 billion in France by 2031 and employs almost 1000 people (AFP 2023a).

This overturning can be explained by all standards that control personal data protection imperatives that must be reconciled with those relating to safeguarding the patient's interests or even public health imperatives. The Covid-19 pandemic created a worrying environment. Digital technology has contributed to this atmosphere of tension and uncertainty by increasing mistrust with regard to the devices implemented by the government, in particular the StopCovid app, which subsequently became Tous Anti-Covid. This application conditioned access to several public places by means of a vaccine pass, proof that users were vaccinated and not carrying Covid. Conversely, a state of vaccination that was not up to date or a positive antigen test which declared that the user was carrying Covid, that the person should be confined or did not have access to certain places such as restaurants, bars or even cinemas for people who refused to be vaccinated. Other than concern regarding the respect of users' freedom, there again we can see a "functional imperative" applied to a context of a health crisis:

> [The use of Stop Covid] had further increased with the installation of the health pass in the summer of 2021, then the vaccine pass at the start of 2022, at the same time amplifying criticism of a tool accused from the start of infringing on individual freedoms. Nevertheless, the ministry confirms that "the initial objectives of the application have been fulfilled" and adds that its users could "still access [their] booklet and store [their] test and vaccination certificates" (AFP 2023b).

The pandemic threw up several questions about the protection, collection and management of health data. The public interest mission defined by the GDPR (art. 9) allows processing of personal health data without individuals' consent. This is one of the rare exceptions which safeguards the essential interests of the person or obeys a legal obligation[21]. This concern did not slow down the partnerships between actors in digital technologies and "traditional" actors in health, and this did not spoil the enthusiasm for connected objects, rather to the contrary. Structuring of digital health requires both legal and organizational assistance. Deploying an app or a platform mobilizes several areas of expertise. First of all, the expertise of the health professional to evaluate the added value of innovation for the pathology concerned, expertise of the DPO (data protection officer) who is in charge of GDPR compliance, and more broadly the protection of personal data (e.g. providing a reminder of the importance of consent for a remote monitoring device[22]), expertise of the information technology security manager to guarantee data output, and expertise of a lawyer who will summarize all of these means of deployment in the context of a contract. Making connections between companies in digital technologies and health establishments requires complex organization to find the delicate balance between patient security, the improvement of working conditions for health professionals and statutory obligations contributing to data protection for these professionals.

21 "It (the CNIL) has specified that 'sending the messages required for the objective set out in article L.33-1 of French laws regarding electronic stations and communications, in the context of stopping the spread of coronavirus (Covid-19), clearly falls within this context'" (Quémémer et al. 2020, p. 65).

22 "Most people are not aware of the methods that are in place to process their data, or how to exercise their rights and also who to contact to do so. This situation is not aligned with the provisions of articles 13 and 14 of the GDPR which state there is an obligation for the people concerned to be informed" (Benayahia 2022, pp. 95–96).

4.3. Prospects: the humanities and social sciences remain too absent from the assessment of connected devices in healthcare

Humanities and social sciences representatives are still missing in decision-making processes. However, several disciplines such as sociology, law or even information and communication sciences have rapidly commented on the obligation for ethical design to take up a place in co-design, for example, with patients[23]. Most research work in the humanities and social sciences casts a critical view which would provide undeniable added value to government roadmaps. This is translated into a difficulty in defining innovation, the extent of connected devices in healthcare, in addition to actors who make up the digital health environment in France (third places, start-ups, for example). Digital health and the difficulties involved in assessing what it involves reveal a gap between institutions and actors in digital health due to the temporal gap between, for example, the decision-making time in a hospital and the obsolescence of an app following its release to the market. Digital health is a multifaceted subject which requires a cross-disciplinary approach. The humanities and social sciences would complement the expertise of health professionals and digital technology professionals in making decisions to deploy a connected device.

23 "Within the humanities and social sciences, scientific literature in management sciences, in particular in management of innovation and in marketing, has a focus on developing a better understanding of connected objects in their functions, for example by drawing up a taxonomy of connected objects which aim to provide recommendations to actors in the market to adapt the offer of connected objects intended for the wider public" (Arruabarrena 2022).

4.4. References

Arruabarrena, B. (2022). Objets connectés : penser les enjeux des technologies connectées sous l'angle de la médiation infocommunicationnelle. *Tic & Société*, 15(2–3), 9–35.

Benayahia, N. (2022). La protection numérique des données de santé. *Journal du droit de la santé et de l'assurance-maladie* (JDSAM), 2022/1(31), 95–96.

Cambon, L. (2016). Le nudge en prévention, troisième voie ou sortie de route. *Santé publique*, 2016/1(28), 43–48.

Catoir Brisson, M.-J., Wrembel, E., Riccio, P.-M., (2022). Objets connectés : enjeux technologiques, enjeux de société. Pour une approche nouvelle de la conception de dispositifs pour le suivi de maladies chroniques. *Tic & Société*, 15(2–3), 161–192.

Dagiral, E., Licoppe, C., Martin, O., Pharabod, A.-S. (2019). Le Quantified Self en question(s). Un état des lieux des travaux de sciences sociales consacrés à l'automesure des individus. *Réseaux*, 2019/4(216), 17–54.

Do Quang, L. (2023). Les objets connectés semblent peu fiables pour étudier le sommeil des personnes qui s'en plaignent. *Tic Pharma*, 8 December [Online]. Available at: https://www.ticpharma.com/story?ID=2450 [Accessed 28 February 2024].

Joly, P.B. (2015). Le régime des promesses technoscientifiques. In *Pourquoi tant de promesses*, Audétat, M. (ed.). Hermann, Paris, 31–48.

Menissier, T. (2016). Innovation et histoire : une critique philosophique. *Quaderni*, 91, 47.

Parizeau, M.-H. (2023). La santé numérique et ses enjeux éthiques : du paternalisme aux normes sociales de santé. *Éthique publique*, 25(1) [Online]. Available at: http://journals.openedition.org/ethiquepublique/7988.

Quémémer, M., Dalle, F., Wierre, C. (2020). *Quels droits face au innovations numériques ?* Éditions Gualini, Paris.

Vitali-Rosati, M. (2024). *Éloge du bug : être libre à l'époque du numérique*. Éditions Zones, Lisieux.

Online resources

AFP (2023a). Cloud : AWS assure que les États-Unis n'auront pas accès aux données européennes. *Le Figaro*, 28 February [Online]. Available at: https://www.lefigaro.fr/flash-eco/cloud-aws-assure-que-les-etats-unis-n-auront-pas-acces-aux-donnees-europeennes-20230228.

AFP (2023b). Covid-19 : le gouvernement met l'appli TousAntiCovid "en pause" à partir de vendredi. *France Info*, 29 June [Online]. Available at: https://www.francetvinfo.fr/sante/maladie/coronavirus/covid-19-le-gouvernement-met-l-appli-tousanticovid-en-pause-a-partir-de-vendredi_5920376.html [Accessed 3 March 2024].

AP-HP (2022a). Future4care, @Hôtel-Dieu et PariSanté Campus annoncent un partenariat inédit au service de la santé numérique. Press Release, 9 March [Online]. Available at: https://www.aphp.fr/contenu/future4care-hotel-dieu-et-parisante-campus-annoncent-un-partenariat-inedit-au-service-de-la [Accessed 25 February 2024].

AP-HP (2022b). Page de présentation de l'étude Evalexplo (actuellement terminée) [Online]. Available at: https://www.aphp.fr/registre-des-essais-cliniques/evaluation-de-la-fiabilite-metrologique-dobjets-connectes-disponibles [Accessed 3 March 2024].

Gouvernement (2024). France 2030, 2 ans de la stratégie "santé numérique" : faire de la France un leader de la santé numérique. Press kit, 18 January.

Gouvernement and CSIC (2021). Innovation Santé 2030 : faire de la France la 1re nation européenne innovante et souveraine en santé. 29 June [Online]. Available at: https://www.gouvernement.fr/sites/default/files/contenu/piece-jointe/2021/06/sante_innov30_a4_07_vdefdp.pdf.

HCL (2024a). La plateforme PLATINES labellisée "tiers-lieu d'expérimentation" en santé numérique. Press Release, 1 February. [Online]. Available at: https://teamhcl.chu-lyon.fr/la-plateforme-platines-labellisee-tiers-lieu-dexperimentation-en-sante-numerique [Accessed 25 February 2024].

HCL (2024b). PLATINES : PLATeforme INnovation Environnement et Santé [Online]. Available at: https://teamhcl.chu-lyon.fr/platines [Accessed 25 February 2024].

Lindermeyer, C. and d'Ortho, M.-P. (2019). L'introduction des objets connectés en medicine. *Hermès, La Revue*, 3(85), 41–42.

Secrétariat général pour l'Investissement (2022). France 2030 : ouverture de l'appel à projets Tiers-Lieux d'expérimentation. 24 February [Online]. Available at: https://www.gouvernement.fr/actualite/france-2030-l-ouverture-de-l-appel-a-projets-tiers-lieux-d-experimentation [Accessed 25 February 2024].

5

Medical Care and Data: The Example of Nanomedicine in Europe

Within the field of advanced technologies applied to health, over the last 20 years, nanomedicine has become a prominent topic, thanks to the promise of products and processes generated by innovation in nanotechnologies. It has created and fueled hope for an improvement in individual and social life through more effective treatment of health pathologies. For this hope to become reality, it was thought in the early 2000s that it was necessary to promote a translation in the field of nanomedicine, where this translation drew inspiration from the medical field itself, referring to the need to improve communication about research into medicine alongside the work of doctors with their patients to reduce latency in the journey taken by scientific innovations towards their practical application. To understand how this project has evolved over the last 20 years and its current impact, we conducted an investigation primarily focused on nanomedicine in Europe, the details of which have published elsewhere (Papilloud and Schultze 2023). In this chapter, we recap the main lessons drawn from this investigation, developed in the context of a macrorelational sociological approach based on socially

---

Chapter written by Christian PAPILLOUD.

legitimized social relationship orders. One of its key concepts, satellization, is mentioned in the context of this chapter (Papilloud and Schultze 2022).

Our chapter begins with the observation that translation in the field of nanomedicine is part of a broader movement to maximize the value of nanotechnologies across all spheres of society, spearheaded by countries that use their influence to coordinate cohesion between science and the economy, thereby promoting technological innovation. This enterprise, which we describe as a satellization of the economy by politics, uses science to complete this satellization and thus develop its influence on both the economy and science. However, it has encountered many difficulties, of which we will mention the most emblematic. Nevertheless, far from marking a stopping point on this satellization process, these difficulties are going to contribute to redefining it and changing the role that nanomedicine plays in it. Initially envisioned as a source of new drugs and treatments to care for patients affected by serious pathologies, nanomedicine has become a framework within which the relationship between medicine and disease can be reconsidered based on the accumulation of data concerning a wide variety of pathologies, where patients are considered active custodians.

Since then, and although satellization continues, political bodies do not seek to promote nanomedicine as such. They work based on an integration model of different technologies – including some which have come from progress made in nanomedicine – in the form of true relational devices capable of registering changes in health pathologies throughout the lives of as many actors as possible. This is the objective related to the development of key enabling technologies (KET) that we will focus on at the end of our contribution. With KETs and the extension of satellization to digital media and to major bodies which bring them into widespread use, it is not only a case of ensuring progress in medical research, but also of achieving the means to take action within these populations of actors; either directly through the intermediary of clinicians, or indirectly, for example, via health insurance providers to develop the movement of actors in society on an individual level, whether or not they declare these pathologies given the context in which they live. This is how the social issues of a digital health ecosystem present themselves when expressed in the context of precision medicine, that the inherent difficulties of promoting nanomedicine contribute to defining as a horizon towards which contemporary medicine must move.

5.1. Satellization or influence by alliance

One of the advantages of the macrorelational sociological approach that we propose and which underpins the theory of the relationship, whose framework we have detailed in other works (Papilloud and Schultze 2022), is to show how specific objects – such as in our case translation in nanomedicine – are structured by strategies of influences between actors and the authorities which represent them to impose a social order. In this case, it is a social order embedding health in technological innovation (Papilloud and Schultze 2022, pp. 399–401). This works best when the information on nanomedicine remains largely within the realm of the initiated, specialists of the subject in scientific and economic fields. This testifies implicitly to the difficulty the media sphere, in our case the written press, faces in finding sufficient attractiveness in this hybrid object, nanomedicine, to justify the work involved in giving it public visibility. Indeed, coverage of nanomedicine in the European press, and internationally, has never exceeded, over the last 20 years (1998–2021), the level it had achieved during the first half of the 21st century[1]. Therefore, it is not surprising to note that

1 See Arnaldi (2008) and https://www.wilsoncenter.org/article/nanotechnology-doesnt-make-news.

when the press talks about nanomedicine, not only does it fail to fully express the complexity of this subject, but in addition, it tends to support the attractiveness of the representation given by political bodies (Kanheman and Tversky 1984; Innis 1991; Nisbet 2002). This reinforces the position of these bodies with regard to nanomedicine, as much as it places them in a position of strength within the scientific and economic environments, which advocate for their own relationship with nanomedicine.

The strengthening of the power of political bodies to influence science and the economy is first affirmed in the context of political initiatives to promote nanotechnologies, which, in Europe at least, follow attempts in the 1980s to encourage science and economics to work together to establish an economy of knowledge in Europe (Benninghoff and Ramuz 2002, p. 45). Since nanotechnologies are likely to lead to cross-disciplinary applications across various fields in economics and society, they provide a new opportunity to use scientific research activities in this field for the purpose of enhancing economic development through products and services that incorporate nanotechnologies. Relayed abundantly by publications aiming to simplify scientific discoveries which align with the viewpoints of public bodies (see in particular Roco and Bainbridge (2002, 2003)), this agenda, the result of long-term mediation work by public bodies on the representation of this relationship between science and economics which harks back to the 1980s, is presented as an opportunity which could not wait (Fiedeler et al. 2004, p. 67; Bundesministerium für Umwelt 2008, p. 3). The satellization of science for use in the economy or for establishing a translation relationship between science and the economy hardly presents a problem for actors in science and for scientific bodies. The more projects they are promised, involving funds for research, infrastructure and personnel, which then indirectly benefit the movement of actors in the field of medical research, the more they allow themselves to be satellized by public bodies. However, for actors and economic bodies, this satellization leads to considerable difficulties. It places companies and industries in an uncomfortable position, requiring them to invest significant resources in partnerships with scientific bodies, with no prospect of a return on investment. They do not know if such a partnership is truly worthwhile, whether it can guarantee the future developments promised to them without compromising their infrastructure, particularly their research department.

These difficulties encourage economic bodies to contest the agenda of political bodies within the realm of public opinion, in terms of general nanotechnologies (see Palmberg et al. (2009), Toth (2009) and Howard and Wetter (2018)). The figures for the growth of a so-called nanotechnology market are the stuff of fantasists, as is the idea that such a market could really exist. The same is true for the indirect benefits of nanotechnologies, particularly job creation. Similarly, it is challenging to develop indicators that accurately measure the actual impact of these technologies on company growth. These critical points of view are confirmed mainly by the fact that public debates on nanotechnologies, organized at the initiative or with the support of political bodies, do not have the expected impact in terms of their satellization strategy being promoted (Werwatz et al. 2006, 2007; Laurent 2010; Papilloud 2010). Here is where this satellization strategy appears to lose ground and weaken, as nanomedicine increases in importance for political bodies.

5.2. Nanomedicine, the challenge of renewed influence

There are three main reasons why nanomedicine has advantages in terms of the satellization of the economy through the satellization of science that political bodies are attempting to set up in this field. On the one hand, medicine is traditionally partial to new technologies arising from research. On the other hand, medical research in Europe is highly regulated, such that the guarantees of a controlled process translating new scientific technologies into the economy and for patients appear to be present (Nel et al. 2010). Therefore, nanomedicine poses fewer problems in terms of discrepancies in the translation process between science and the economy, as envisioned by political bodies. Additionally, it offers advantages in terms of the image of nanotechnologies among the general public, particularly because the primary objective of nanomedicine is to combat serious illnesses and to promote health (Bottini et al. 2011). Although it was beginning to fade, the satellization implemented by political bodies based on nanotechnologies found a source of renewed energy in nanomedicine, even a source of strength, given that nanomedicine appears to present a context that brings together all the conditions for a successful translation of science into the economy (ETPN 2005).

Nevertheless, if we examine the case of nanomedicine in Europe in more detail, it is clear that optimism is short-lived, because this renewed satellization throws up issues right from the outset, in other words, as soon as political bodies implement formalization work. For example, the organization of conferences, publications on nanotechnologies and medicine, meetings with project managers, and canvassing of research funding bodies who may structure their own satellization to prepare the context for the meeting between actors and bodies in scientific and economic environments. It appears that the expected translation in nanomedicine is taking longer than anticipated, specifically from 10 to 16 years if everything proceeds as planned (Bundesministerium für Bildung und Forschung 2009, p. 28; Nanomedicine 2009, p. 21), which is sufficient to temper investors' enthusiasm. In fact, while the European Commission and its national representatives, in particular British, German and French, rely heavily on the mobilization of large companies and health industries that are behind European nanomedicine and on their broad support, these companies and industries, in particular large pharmaceutical industries, are announcing their withdrawal from the sector. They do not wish to invest in European nanomedicine at the moment, as its development in Europe is still too uncertain to guarantee a return on investment. The financial risks are quite simply too high. This is a significant setback for the satellization strategy of European political bodies, which is accompanied by high costs in terms of work and time, which must now be invested to try, come what may, to complete their satellization.

Concretely and since political bodies cannot rely on major pharmaceutical industries to satellize other economic actors, they must identify such actors themselves to win them over (Eaton 2007; Eaton et al. 2015). These actors should already be active in the health market. Additionally, they must be open to partnerships with actors from the scientific community to collaborate on developing products and services that use nanotechnologies developed for medical applications. A second, equally important, difficulty arises alongside this first issue. Not only must the economic actors be identified and persuaded of the advantages of a translation in nanomedicine, but also concretizing partnerships envisaged with scientific actors and bodies requires research and work to identify suitable actors and bodies that are prepared to support the satellization supported by European political bodies. Long-term mobilization work of this type will necessarily generate a certain number of questions within the very bodies that the European Commission creates specifically to support European nanomedicine and its translation to nanomedicine, first in line being the European Technology Platform on Nanomedicine (ETPN) established in 2005.

5.3. Doubts and decisions

The ETPN is a forum for nanomedicine and a branch of the European Technology Platform (ETP). It is composed of working groups that implement the European Commission's roadmap, drawn up to develop nanomedicine in Europe. The primary objective is to fulfil the promises made by nanomedicine in various fields of medicine and the pharmaceutical sector for the benefit of patients. This project is therefore supported directly by the satellization implemented by European political bodies, with the ETPN taking into account that only the framework of a fruitful collaboration between science and the economy will allow the promises made by nanomedicine to come to pass (see in particular ETPN (2005)). However, implementation of this framework raises many questions, which not only relate to how to establish and promote partnerships of this type, but also to the very foundation of satellization developed by European political bodies. Given the difficulties encountered by this satellization described above, there is a growing doubt about whether political bodies are genuinely able to materialize this partnership between science and economics in the field of nanomedicine (see Papilloud and Schultze (2023)). In addition, supposing they achieve this, can such a partnership really not only lead to products and services from nanomedicine, but also to products and services which reach the health market within a reasonable time and without the process becoming too complex? Should we not switch to a new strategy here? Should we not instead encourage the economy by providing it with the tools to initiate such partnerships based on products and services that it considers promising, given the market's maturity and expected market shares? Suppose we consider the problem from a financial perspective. In that case, more substantial support for the economy would reduce pressure on political bodies, which, otherwise, would need to use public money – taxpayers' money and therefore potentially patients' money –

to fund such partnerships. Moreover, what do we observe when we examine the fallout of the projects funded by this public money? We observe that they certainly lead to patents, but this very rarely goes any further, or, to express this differently, the impact of these projects rarely reaches the patients.

Such questions circulate both in the ETPN's circle and between its members and its founders, showing the extent of the combat engaged in between the representatives of the various political bodies and the various affiliated services in European countries (ministries of research, ministries of education, organizations promoting the economy, organizations promoting scientific and technical culture, for example). Those on one side wish to promote the satellization of science in order to satellize the economy. Those on the other side argue that this strategy has yielded few concrete results and that a different approach should be considered. To ward off the threat of division among ETPN members, a compromise solution is proposed, consisting of setting up, under ETPN's control, a body responsible for locating, mobilizing and selecting actors in science and the economy (Gouze and Boisseau 2013). The objective is to identify these actors to promote projects that not only bring to life the desired partnership between science and the economy, but also ensure its success, which is then measured by the capacity of these partnerships to introduce innovative products and services into the healthcare market (Eaton 2011). Although some members of the ETPN view this initiative from a critical distance, it can still take the upper hand. This leads to the creation of a body for mediation within the ETPN, in the form of the Translation Advisory Board (TAB), which has some analogous examples in medicine (Hodgson et al. 2011; Pratt et al. 2012). It is a group of experts responsible for determining the potential of nanomedicine projects in Europe, advising and supporting promising projects, and facilitating the translation of science into the economy, from idea to product. The TAB is therefore a tool which supports the formalization work of political bodies. It is open to actors in science and the broader economy. However, it must dedicate particular attention to small- and medium-sized enterprises (SMEs), which, due to the withdrawal of large pharmaceutical industries that support European nanomedicine, must now bear the risk.

The hope of inducing changes in the satellization of European political bodies, their correction and therefore reinforcement where necessary, is based on the TAB (ENATRANS 2018). However, from the first meetings that the TAB set up with potential project leaders to evaluate their projects and decide on support, it has been a disappointment. It is clearly the case that these projects are not sufficiently mature for the experts in the TAB to begin their intended work to assist these projects, as well as the ETPN and the European Commission. Indeed, the initiative is not a failure, far from it. For example, scientific actors contact the TAB in large numbers. Nevertheless, most often, they misuse its purpose, particularly the project assessment procedure implemented by the TAB, which encourages these actors to test their ideas and concepts with experts from the TAB before withdrawing their projects back to their laboratories. Fewer economic actors contact the TAB, and although they submit their project, this is most often because they are seeking alternative perspectives or other fields of application related to the technology they wish to develop. Over the course of the meetings, it also became apparent that the indirect objective of encouraging SMEs to work in relation to nanomedicine is not realistic, due to the lack of projects and candidates from SMEs. In other words, while it remains focused on nanomedicine, the TAB loses its purpose, its *raison d'être*, which is why the TAB will be opening up to projects in areas of health technology that mostly have nothing to do with nanomedicine. This redirection of the TAB's work marks a new redirection of satellization, carried forward by European political bodies, which is reflected in the marginalization of support for nanomedicine and the emergence of a new subject of promotion: precision medicine.

5.4. From nanomedicine to precision medicine

TAB experts understand that their initial focus on nanomedicine and translation to nanomedicine does not allow them to fulfil their mission. They must therefore reorient themselves, consider other projects that will produce results more quickly and identify the corresponding actors who are easier to mobilize and work at the interface between technology and health. For the TAB experts, this reconsideration of their mission is not a questioning, but rather an adjustment of a strategy intended to generate new ideas and concepts that could encourage, through means, satellization of political bodies that would otherwise stall. This redirection is ratified by the simultaneous approval of two documents, one of which is the summary report produced by the TAB in the context of the European coordination project NOBEL (Spring et al. 2020). The other is a white paper from the German platform NanoBioMedizin, which played a crucial role both within the ETPN and, with regard to the TAB experts, in coordinating the work of these two bodies in support of nanomedicine in Europe (Deutsche Plattform NanoBioMedizin 2020).

In these two documents, nanomedicine is only marginally featured when mentioned at all – the term nanomedicine is, for example, absent from the NOBEL summary report. On the other hand, other media, mainly from the context of information and communication technologies (ICTs) and from computer science, are brought to the forefront. They should be more involved and integrated into medical practice to promote what the NOBEL summary report refers to as a continuum of care for patients throughout their lifetime (Spring et al. 2020, p. 6). This expression evokes the implementation of a relational device with the patient themselves, which becomes the carrier of various miniaturized information systems – at the surface of the body or within the body – which register and provide information about the occurrence and evolution of pathologies. In the long term, the objective would be to draw informed conclusions from these data about the onset of these pathologies and variations that occurred throughout the patients' lifetimes, to direct patients towards the corresponding social circulation and social bodies appropriate for their state of health. In other words, the relational devices envisaged must, on paper, participate in managing the circulation of social actors – and consequently the corresponding preparation of social bodies which support them – depending on the weight and gravity of the pathologies they declare and with which they live in a latent or declared state throughout their lifetime.

This new orientation of satellization set up by political bodies at a European level concludes the modest degree of development of nanomedicine and its importance for medicine as a whole. As stated by Zingg and Fischer, if we focus on the progress made in the field of nanopharmaceuticals or observe the development of patents in nanomedicine, we observe consolidation rather than expansion over the last two decades (Zingg and Fischer 2019, p. 2). In addition, the authors note that "innovations relate primarily to the control of uncertainties that are inherent to nanomedicine", in other words, controlling the risks to health induced by nanotherapies "based on tried and tested technologies". In contrast, companies and industries "have typically preferred fields for which the scientific groundwork has already been done, where other investors have already cleared the way for nanomedicine" (Zingg and Fischer 2019). In other words, the consolidation of existing research and companies' conservative investment strategies has led to the development of nanotechnology products focusing on established markets, resulting in a withdrawal of investment from nanomedicine and the underutilization of its innovative potential (Zingg and Fischer 2019).

A similar observation can be drawn from Noury and Lafontaine's investigation on nanomedicine in France and Canada. Nanomedicine indirectly concerns the relationship between medicine and patients, because their fundamental objective is to "reshape the body, which implies controlling and refiguring vital processes and generating a new biological normality produced technically" from individuals' molecular properties (Noury and Lafontaine 2014). Consequently, the purpose of nanomedicine is not "to restore normal operation of these processes", in other words to provide care for the patient: "Living things are fundamentally considered to be a subject which can be summarized, reproduced, manipulated or transformed, without, however, altering its underlying nature given that this is defined by its artificiality, its plasticity" (Noury and Lafontaine 2014). Or as stated by Baptista: "Perhaps the idea of personalizing medicine is still nothing more than a vision, and the real concept is reinforcing and optimizing the precision of current nanomedicines" to overcome the difficulties of a transfer in nanomedicine and redefinition of the role of nanomedicine (Baptista 2016, p. 228). Redefining the role of nanomedicine means leaving behind the idea of "one drug fits all" that it conveys, which has proven to be neither scientifically pertinent nor economically beneficial for drug manufacturers, and from a health point of view for patients. Nevertheless, "[...] nanotechnology-enabled approaches have demonstrated astonishing potential to transform medicine", which is why more work is required on the precision of nanomedicine, in order to set up "smoother transitions from the 'one drug to fit all' state to highly personalized treatment modalities" (Herrmann and Rösslein 2016, p. 2).

Therefore, we should no longer see nanomedicine in terms of curing patients, but instead as a springboard among other technological innovations towards redefining medicine as precision medicine. It is a redefinition which, according to Hermann, is inevitably based on the development of theranostic strategies, in other words, treatment and diagnosis in relation to the collection of data on pathologies mentioned above (Herrmann and Rösslein 2016). A suitable ecosystem for the development of this precision medicine is key enabling technologies (KET) which are today becoming the new fuel for the satellization of European political bodies.

5.5. KETs, a new challenge for satellization

KETs and in particular KET 4.0s are associated with six technologies linked primarily to ICT in the field of health, in other words:

– advanced additive manufacturing processes, such as those used in autonomous systems or sensor technology,

– advanced nanomaterials such as biomaterials,

– life sciences technologies such as nanotechnologies, neurotechnologies and bio-engineering;

– micro/nanoelectronics and photonics for the manufacture of integrated circuits, quantum computers, IoT sensors (Internet of Things or information systems connected in a network and implemented in objects to create connected objects) capable of significantly increasing the calculation power of computers,

– artificial intelligence for data processing, robotics, implementation of autonomous services and systems,

– security and connectivity technologies such as cryptography, networks and IoT protocols ((European Parliament 2021, p. 5); see also (High Level Group 2018, pp. 21–23)).

KETs are becoming the central pillar of the new direction taken by the satellization strategy developed by political bodies in Europe. This new direction owes nothing to chance; it responds to the requirement that political bodies have always associated with their satellization, which is to reach the largest number of people and convince all involved of the attractiveness of nanotechnologies and nanomedicine. Since for these bodies it is a case of representation, in other words, of themselves as representatives of everyone and in our particular context of the health of everyone, that nanomedicine has become the new guardian of. In the same way as for nanotechnologies, we have also been able to demonstrate the extent to which these bodies, whether European or national, have mobilized the public by involving various media and offering different formats of participation in support of their satellization strategy (Papilloud 2010, 2016). This new direction taken in the satellization of political bodies, therefore, corresponds to an additional iteration where, assisted by bad experiences, it is no longer a case of achieving direct support from the population. It is more a case of integrating the population indirectly in this satellization through digital media which measure and report on the state of health of individuals in the form of data. Here, theranostics plays a primary role, as it represents the practical context that facilitates the integration of these various media for the purpose of recording and collecting data.

The European Commission had understood this and emphasizes it in its Industrial Policy (see European Commission (2010, 2012a, 2012b)), the primary pillar of the "coordinated European Strategy for KETs" ((High Level Expert Group on Key Enabling Technologies 2015, p. 34); also see (European Commission 2009, 2012a)), which aims to "force the different healthcare industries, Pharma, MedTech and IVD to co-develop diagnostic and therapeutic systems in a cross-technology and cross-industry approach"

(NOBEL 2020, p. 11). After concentrating on science and the economy, public bodies are extending their satellization to the digital media commonly used in the medical field and implemented around, on or within patients. An in-depth examination of the documents associated with the NOBEL project reveals what the initial steps of this new satellization must entail.

We have stated in the above that the NOBEL summary report does not mention the term nanomedicine as such. However, it does reference nanotechnologies and their health applications (see Spring et al. (2020, pp. 16–22)). Similarly, regarding the continuum of care, the NOBEL summary report only mentions the idea of a transfer as a marginal point, in the context of technologies providing patient assistance for therapeutic purposes (Spring et al. 2020, p. 11). This relativization of the importance of nanomedicine and the translation in nanomedicine help to more easily highlight the key point of the European innovation policy's agenda: KETs and KET 4.0 (High Level Group 2018, p. 19). In the NOBEL summary report, KETs are encapsulated in the notion of HealthTech around which the "European ecosystem of nanobiomedicine" should be mobilized to "develop a strategic vision for HealthTech in Europe" (see: https://cordis.europa.eu/project/id/766492/reporting) (Spring et al. 2020).

According to NOBEL, the term HealthTech refers to "any technology, device or digital solution designed to improve the health of patients" (see: https://nobel-project.eu/technologies-in-medtech/what-is-healthtech/). Such technologies must, on the one hand, contribute to the development of medical devices [...] which are intended to have a therapeutic and diagnostic effect on people by physical means. Additionally, these technologies must also serve to develop *in vitro* diagnostic systems in support of non-invasive tests, which allow biomarkers of pathologies in biological samples to be analyzed. Lastly, HealthTech must enable the implementation of digital health solutions (*ibid.*). Technology sectors which must contribute to the achievement of this objective are:

– nanomedicine;

– photonics;

– robotics;

– biomaterials known as advanced biological materials;

– digital health (*ibid.*).

Let us examine in more detail how these five technological sectors can contribute to healthcare based on KETs and the implications of this for the new orientation of satellization developed by public bodies.

Photonics, biomaterials and nanomedicine are not made available exclusively for nanobiomedical theranostics. They must also contribute indirectly to the entry of data collected from a large number of patients (big data). Thus, the use of biomaterials is planned to support research "against pathologies through real time monitoring of physiopathology" (see: https://nobel-project.eu/technologies-in-medtech/advanced-materials-for-healthcare/). Photonics must allow "continuous remote monitoring of patient health parameters" in addition to "massive data transfer" (see: https://nobel-project.eu/technologies-in-medtech/photonics-for-healthcare/). Nanomedicine participates in the widespread data integration of the health system: "nanocomponents included in wearable devices" must allow "remote monitoring of diseases & conditions, source of patient clinical data" (see: https://nobel-project.eu/technologies-in-medtech/nanotechnologies-for-healthcare/). In this same context, robotics as a "branch of engineering and science that deals with the design, manufacturing and use of mechanical and virtual robots" must allow "sensory feedback" to be recorded and participate in "information processing" (see: https://nobel-project.eu/technologies-in-medtech/robotics-for-healthcare/). Lastly, digital health must lead to the implementation of "digital sciences for healthcare" responsible for analyzing "the use of Information and Communication Technologies (ICT) products, services and processes" in health "combined with organisational changes in healthcare" (see: https://nobel-project.eu/technologies-in-medtech/digital-health/). Digital health must lead to medicine that is "more preventive, personalised, precise and patient tailored" while contributing to a strong increase in the efficiency of the health system across society (*ibid.*). These objectives must be reached by four different means, which are by "the use of data generated by medical devices", by "progresses of machine-learning and artificial intelligence, coupled with the big data gathered from genomics and proteomics", via "super-computers from IBM or Google which can diagnose efficiently complex pathologies [...] thanks to the ability of AI to compare clinical data of the patient with a huge clinical database" and through the integration of artificial intelligence in "new smart medical devices" which are likely to recover large quantities of data "from the patient level to the population level" (*ibid.*).

With digital health, the redirection of the satellization undertaken by European political bodies, initially based on the promotion of nanomedicine, and now on KETs, remains underpinned by the topic of the collaborations between science and the economy, with the addition of digital media and its respective bodies. This envisioned collaboration between the economy and the sphere of big-tech increases in importance as the integration of media in the satellization practiced by political bodies complexifies this satellization and opens it up to more uncertainties. However, for political bodies, this potential vulnerability will very quickly be surpassed because this "vision for HealthTech" is within reach, and the conditions required for it to come to fruition have already been met. As was the case with the satellization of science and economics by political bodies in the context of nanotechnologies, then in nanomedicine, a comparable rhetoric of a done deal is established, supporting this new direction in the satellization of European political bodies and which must eliminate all doubts with regard to its outcome.

In this context, we mention Siemens Healthineers and General Electrics Healthcare, two notable representatives of the convergence of the economy and the media sphere in the health sector. As leaders in the market of large medical devices, these two large industrial groups collect data on patients using their digital facilities. They register them in large databases, allowing them to evaluate them using artificial intelligence. Their objective is to create patient-focused innovations which should in turn encourage implementation of a "worldwide digital ecosystem […] with a broad offer of services in the field of medical information technologies" ((Huss 2019, p. 34); see also (General Electric Healthcare/University of California 2016; Berghold 2022, pp. 12-13)). These two initiatives should inspire the creation of companies which, for example, count on cloud solutions and wearables, in other words portable measurement devices, smart devices which can range from smart watches to "diagnostic and partially implantable nanoparticles", or even to "contact lenses which measure glucose" (Huss 2019, p. 33). Using this incubator of future economic actors, it will be possible to exploit "the largest possible quantity of data for an entire lifetime" for the largest possible number of people, and for a global and digital health system to be implemented using this data articulated around artificial intelligence (*ibid.*, p. 34). Large companies are already following this route, such as the parent company Google Alphabet, which includes, besides Google, the subsidiaries Calico, DeepMind, Google Fiber, X and Verily Life Science (formerly

Google Life Science) ((Berghold 2022, pp. 12–13), see also: https://abc.xyz, https://www.calicolabs.com/, https://www.deepmind.com/, https://x.company, https://verily.com/), the division Watson Health at IBM (renamed Merative) in collaboration with the Machine Learning Platform at Apple (see: https://www.ibm.com/de-de/watson-health) and the company Philips HealthCare in collaboration with Amazon Web Service (see: https://www.philips.de/healthcare, https://aws.amazon.com/de/solutions/case-studies/philips/).

As we can see in these few examples, the issue of this rhetoric intended to legitimize redirection of the satellization deployed by political bodies towards science, the economy and the media sphere, is indeed based on this new theme of alliance between the economy and the media sphere, which now carries all hopes related to this new satellization. These hopes no longer prioritize treatment of patients, an indirect challenge of this satellization, in the same way as precision medicine, which should allow their care to be improved, in other words, with a higher success rate and less invasively. The direct challenge is the implementation of a health system supported by the KETs, which monitor pathologies in real time at the scale of human populations in order to create databases and analyses corresponding to these data. It will then be possible to develop a dedicated digital health ecosystem where scientists, pharmaceutical companies, tech giants, public and private health insurance providers work in concert on the composition of life trajectories of individuals, depending on whether they declare a given pathology or not and depending on the time for which they must live with this pathology. The redirection of the satellization of political bodies, following the failure of the satellization used to promote nanomedicine in Europe, reflects the dream it has distant hopes for, which is currently developing.

5.6. Conclusion

In this chapter, we have seen how, based on the failure to promote nanomedicine by encouraging science and economics to work together, European political bodies have changed direction in their strategy of influence, on the one hand, by changing the objective of this strategy and, on the other hand, by extending its reach to integrate the media and large digital technology companies which produce them. While political bodies have focused on nanomedicine and a transfer in nanomedicine as an extension of promoting nanotechnologies in all fields of society, they have repeatedly faced the difficulty of establishing a partnership between science and economics in this field, in addition to the indifference, if not resistance, of the public.

Reorientation of their satellization of science and, through this, of economics has taken place progressively, counting on technologies indirectly associated with the treatment of patients such as KETs. Under the principle of digital health, of which the by-product is precision medicine, the true objective of this focus on KETs is, however, to redefine the health system and its actors in science, economics and the media based around management of the life pathway of social actors on the basis of the pathologies they declare. Underpinned by discourse praising a future of precise, more effective, less invasive and more personalized medicine, a new definition of individual health emerges based on the analysis of data drawn from pathologies and their changes in real time within the population. This is the direction set by the new satellization implemented by the political bodies, now extended to tech giants by which these bodies hope it will give a new lease of life to the partnership between science and the economy in the field of health, where nanotechnologies and nanomedicine have never really achieved an enthusiastic reception.

5.7. References

Arnaldi, S. (2008). Converging technologies in the Italian daily press 2002–2006: Preliminary results of an ongoing research project. *Innovation – The European Journal of Social Science Research*, 21(1), 87–94.

Baptista, P. (2016). Precision nanomedicine in cancer: How far are we from personalization? *Expert Review of Precision Medicine and Drug Development*, 1(3), 227–228.

Benninghoff, M. and Ramuz, R. (2002). Les transformations des systèmes de recherche publique suisse et français. *Politiques et management public*, 20(1), 31–47.

Berghold, A. (2022). Tech-Gigaten im Gesundheitswesen. Report, Bertelsmann Stiftung, Gütersloh [Online]. Available at: https://www.bertelsmann-stiftung.de/fileadmin/files/user_upload/VV_Tech-Giganten_im_Gesundheitswesen1.pdf.

Bottini, M., Rosato, N., Gloria, F., Adanti, S., Corradino, N., Bergamaschi, A., Magrini, A. (2011). Public optimism towards nanomedicine. *International Journal of Nanomedicine*, 6, 3473–3485.

Bundesministerium für Bildung und Forschung (2009). Status Quo der Nanotechnologie in Deutschland. Report, Bundesministerium für Bildung und Forschung.

Bundesministerium für Umwelt, N. und R. (2008). NanoKommission – Zwischenbilanz im Dialog. Ergebnisdokumentation der Zwischenbilanz am 20. Februar 2008 im DBB-Forum in Berlin. Report, Bundesministerium für Umwelt, Naturschutz und Reaktorsicherheit [Online]. Available at: https://www.bmu.de/fileadmin/Daten_BMU/Download_PDF/Nanotechnologie/nano dialog_1_zwischenbilanz.pdf.

Deutsche Plattform NanoBioMedizin (2020). Aktionspapier 2020 NanoBioMedizin. Report, Deutsche Plattform NanoBioMedizin, Frankfurt am Main.

Eaton, M. (2007). Nanomedicine: Industry-wise research. *Nature Materials*, 6, 251–253.

Eaton, M. (2011). How do we develop nanopharmaceuticals under open innovation? *Nanomedicine: Nanotechnology, Biology, and Medicine*, 7, 371–375.

Eaton, M., Levy, L., Fontaine, O. (2015). Delivering nanomedicines to patients: A practical guide. *Nanomedicine: Nanotechnology, Biology, and Medicine*, 11, 983–992.

ENATRANS (2018). Publishable summary. Report, ENATRANS, Brussels, No. Ref. Ares(2018)1262337.

ETPN (2005). European Technology Platform on NanoMedicine. Nanotechnology for Health. Vision Paper, The European Commission, Brussels.

ETPN (2009). Roadmaps in nanomedicine towards 2020. Report, European Technology Platform on Nanomedicine.

ETPN (2011). Improving translation of public healthcare nano-research in Europe. White paper to the horizon 2020 framework programme for research and innovation. Recommendations from the Nanomedicine Community. Report, European Technology Platform on Nanomedicine.

European Commission (2009). Preparing for our future: Developing a common strategy for key enabling technologies in the EU. Report, Commission to the European Parliament, the Council, the European Economic and Social Committee and the Committee of the Religions, Brussels [Online]. Available at: https://eur-lex.europa.eu/legal-content/EN/TXT/PDF/?uri=CELEX:52009DC0512&from=EN.

European Commission (2010). An integrated industrial policy for the globalisation era. Putting competitiveness and sustainability at centre stage. Report, Commission to the European Parliament, the Council, the European Economic and Social Committee and the Committee of the Religions, Brussels [Online]. Available at: https://eur-lex.europa.eu/LexUriServ LexUriServ.do?uri=COM:2010:0614:FIN:EN:PDF.

European Commission (2012a). A European strategy for key enabling technologies – A bridge to growth and jobs. Report, Commission to the European Parliament, the Council, the European Economic and Social Committee and the Committee of the Religions, Brussels [Online]. Available at: https://eur-lex.europa.eu/LexUriServ/LexUriServ.do?uri=COM:2012:0341:FIN:EN:PDF.

European Commission (2012b). A stronger European industry for growth and economic recovery. Industrial Policy Communication Update. Report, Commission to the European Parliament, the Council, the European Economic and Social Committee and the Committee of the Religions, Brussels [Online]. Available at: https://eur-lex.europa.eu/LexUriServ/LexUriServ.do?uri=COM:2012:0582:FIN:EN:PDF.

European Parliament (2021). Key enabling technologies for Europe's technological sovereignty. Report, Panel for the Future of Science and Technology, European Parliament Research Service, Brussels [Online]. Available at: https://www.europarl.europa.eu/RegData/etudes/STUD/2021/697184/EPRS_STU(2021)697184_EN.pdf.

Fiedeler, U., Fleischer, T., Decker, M. (2004). Roadmapping als eine Erweiterung des "Methoden-Werkzeugkastens" der Technikfolgenabschätzung? *Technikfolgenabschätzung – Theorie und Praxis*, 13(2), 65–70.

General Electric Healthcare/University of California (2016). Big data, analytics & artificial intelligence. The Future of Health Care is Here. Report, Panel for the Future of Science and Technology, European Parliament Research Service, San Francisco [Online]. Available at: https://www.gehealthcare.com/static/pulse/uploads/2016/12/GE-Healthcare-White-Paper_FINAL.pdf.

Gouze, N. and Boisseau, P. (2013). Nanomedicine 2020. Contribution of nanomedicine to horizon 2020. White Paper, The European Commission, Brussels.

Herrmann, I.K. and Rösslein, M. (2016). Personalized medicine: The enabling role of nanotechnology. *Nanomedicine*, 11(1), 1–3.

High Level Expert Group on Key Enabling Technologies (2015). KETs: Time to act. Report, Europäische Kommission, Brussels [Online]. Available at: https://www.eusemiconductors.eu/sites/default/files/uploads/2015_HLG_KETs_Final_Report_Final_version.pdf.

High Level Group (2018). Re-finding Industry – Defining Innovation. Report, Europäische Kommission, Brussels [Online]. Available at: https://op.europa.eu/o/opportal-service/download-handler?identifier=28e1c485-476a-11e8-be1d-01aa75ed71a1&format=pdf&language=en&productionSystem=cellar&part=.

Hodgson, P., Towers, A., Keast, D., Kennedy, A., Pritzker, R., Allen, J. (2011). Lymphedema in Canada: A qualitative study to help develop a clinical, research, and education strategy. *Current Oncology*, 18(6), 260–264.

Howard, S. and Wetter, K. (2018). Nanotechnology and geopolitics: There's plenty of room at the top. In *Nanotechnology and Global Sustainability*. Maclurcan, D. and Radywyl, N. (eds). CRC Press, Boca Raton, 69–116.

Huss, R. (2019). *Künstliche Intelligenz, Robotik und Big Data in der Medizin*. VS-Verlag, Wiesbaden.

Innis, H. (1991). *The Bias of Communication*. University of Toronto Press, Toronto.

Kanheman, D. and Tversky, A. (1984). Choices, values and frames. *American Psychologist*, 39(4), 341–350.

Laurent, B. (2010). *Les politiques des nanotechnologies*. Charles Léopold Mayer, Paris.

Nel, A., Grainger, D., Alvarez, P., Badesha, S., Castranova, V., Ferrari, M., Godwin, H., Grodzinski, P., Morris, J., Savage, N., et al. (2010). Nanotechnology environmental, health, and safety issues. In *WTEC Panel Report on Nanotechnology Research Directions for Societal Needs in 2020. Retrospective and Outlook*, Roco, M., Mirkin, C., Hersam, M. (eds). WTEC, 107–156 [Online]. Available at: https://www.nano.gov/sites/default/files/pub_resource/wtec_nano2_report.pdf.

Nisbet, M. and Lewenstein, B. (2002). Biotechnology and the American Media: The policy process and the Elite Press, 1970 to 1999. *Science Communication*, 23(4), 359–391.

NOBEL (2020). Mobilising the European nano-biomedical ecosystem. Report on socio-economic impact. Consolidated version, Deliverable D5.3. Europäische Kommission Report, Brussels [Online]. Available at: https://ec.europa.eu/research/participants/documents/downloadPublic?documentIds=080166e5d5b5c9ae&appId=PPGMS.

Noury, M. and Lafontaine, C. (2014). De la nanomédecine à la nanosanté : vers un nouveau paradigme biomédical. *Socio-anthropologie*, 29, 13–35.

Palmberg, C., Dernis, H., Miguet, C. (2009). *Nanotechnology: An Overview based on Indicators and Statistics*. OECD, Paris.

Papilloud, C. (2010). *Gouverner l'infiniment petit*. L'Harmattan, Paris.

Papilloud, C. (2016). La science en action ? Luttes d'expansion institutionnelle à l'exemple de la nanomédecine en Allemagne. *Terminal*, 119. doi: 10.4000/terminal.1562.

Papilloud, C. and Schultze, E. (2022). *Skizze einer Theorie der Relation*. VS-Verlag, Wiesbaden.

Papilloud, C. and Schultze, E. (2023). *Translation in der Nanomedizin*. VS-Verlag, Wiesbaden.

Pratt, B., Zion, D., Lwin, K., Cheah, P., Nosten, F., Loff, B. (2012). Closing the translation gap for justice requirements in international research. *Journal of Medical Ethics*, 38, 552–558.

Roco, M. and Bainbridge, W. (2002). *Converging Technologies for Improving Human Performance: Nanotechnology, Biotechnology, Information Technology and Cognitive Science*. National Science Foundation, Arlington.

Roco, M. and Bainbridge, W. (2003). *Converging Technologies for Improving Human Performance: Nanotechnology, Biotechnology, Information Technology and Cognitive Science*. Kluwer Academic Publishers, Dordrecht.

Spring, K., Gouze, N., Galvin, P., Gramatica, F., Weltring, K.-M. (2020). A model for a life-long personalised Continuum of Integrated Care revolutionising healthcare delivery: Description of technological impact. Report, NOBEL, Münster.

Toth, E. (2009). Buying the nano-market. *HesaMag*, 1, 14–17.

Werwatz, A., Belitz, H., Kirn, T., Schmidt-Ehmcke, J. (2006). *Innovationsindikator Deutschland 2006*. DIW, Berlin.

Werwatz, A., Belitz, H., Kirn, T., Schmidt-Ehmcke, J. (2007). *Innovationsindikator Deutschland 2007*. DIW, Berlin.

Zingg, R. and Fischer, M. (2019). The consolidation of nanomedicine. *WIREs Nanomedicine and Nanobiotechnology*, 11(6), 1–6.

6

Sociopolitical Challenges of Digital Uses

Examining connected devices in healthcare inevitably leads to considering the broader digitalization of society, a process now so intensive that we can speak of a digitalized society. This transformation has been analyzed since the 1980s (CNIL 1978; Maise and Vitalis 1985; Vitalis 2009), although it was already considered as early as the 1950s–1960s (Mumford 1963; Ellul 1990). The digitalization of the health sector also has a relatively long history (Carré and Lacroix 1999). Connected objects in health are rooted in a contemporary process of acceleration in the development of digital technologies, in other words, not information technology for information system management, but the social appropriation of digital information and communication technologies and the Internet. This appropriation goes beyond the questions of equipment and access and concerns uses and their meanings in the lives of users. In fact, even over a short timeframe, switching from a smartphone connected to the Internet to connected objects presents sociopolitical challenges associated with digital uses[1]. In this chapter, we

Chapter written by Geneviève VIDAL.

1 See the French framework for digital uses ARCEP – ARCOM: https://www.arcep.fr/actualites/actualites-et-communiques/detail/n/usages-numeriques-en-france-240624.html.

propose discussing these challenges in terms of the expected solutions, which raise major issues in society.

To understand this ambivalent situation, the approach is based on more than 20 years of work on the digital uses of information and communication technologies, considering objects related to digital health as positioned within the broader dynamic of digital uses. However, health data are sensitive and require urgent protection (Bévière-Boyer 2021) in the era of connected objects in this sector and the datafication of society's activities through algorithmic processing and artificial intelligence systems (AI).

This situation presents sociopolitical challenges regarding digital uses, including for health professionals. The use of data on digital networks for predictive analyses is accompanied by discourse relating to technological solutionism. Yet, the reality appears to be closer to the use of data for economic and political purposes. The social acceptability of connected objects is part of this movement of belief in digital technologies that can solve many problems regarding resources for public health (equipment, number of patients, doctors and caregivers, appointment times, remuneration, etc.). The acceleration and normalization of ubiquitous monitoring and control devices, which dictate the invasive digital injunction to the point of transforming practices (Vidal and Panico 2019), while not preventing a reflective posture and a certain knowledge of the rights to personal data protection, do not provide a full understanding of the risks. They should not be considered only in terms of ethics[2]. The connectivity of objects is supposed to help manage a large number of problems, or anticipate them in order to

2 The INSERM ethics committee, 2022, see: https://www.inserm.fr/seminaire/donnees-massives-de-sante-enjeux-ethiques-pour-la-recherche/. National ethics consultation committee, concerning health data platforms, February 2023, see: https://www.ccne-ethique.fr/sites/default/files/2023-05/CCNE-CNPEN_GT-PDS_avis_final27032023.pdf.

avoid them. Thus, everything would become smart: health, transport, pollution reduction, etc. But the security of traceability and the contributions of AI – with machine learning and deep learning – are essential for health policy and the health economy, in a context favorable to the development of connected objects in health and e-health[3].

First, there is a discussion on the digitalization of society, which includes the development of connected objects in healthcare. Second, the acceptability and ambivalence of their uses are considered. Finally, the sociopolitical challenges of these uses, which are subject to conditions, are identified.

6.1. The digitalization of society

To identify the acceptability and uses of connected objects in society, the relationship between technical and societal aspects should be kept in mind, in terms of surveillance and prescription, while consent is not entirely clear given such intense acceleration and automation. In the era of the contemporary deployment of artificial intelligence (AI), which provokes enthusiasm and solutionist fervor, the relationship between technical aspects and society is an ambivalent one, with fears of the risks also heightened. Yet, AI renews cybersurveillance to track down profiles and serve security and economic interests. With the automation of its deductions, AI requires massive processing of data (big data) and sophisticated cross-referencing, with machine learning that ISO (see: https://www.iso.org/artificial-intelligence/machine-learning) refers to as "a type of artificial intelligence that allows machines to learn from data without being explicitly programmed. It does this by optimizing model parameters (i.e. internal variables) through calculations, such that the

3 See: https://cnnumerique.fr/files/2017-10/CNNum--rapport-ambition-numerique.pdf (pp. 302–312).

model's behaviour reflects the data or experience [...] enabling the ML model to learn and make predictions or decisions based on data science".

The unavoidable general data protection regulation (GDPR) is of course mentioned, and the demand for ethics appears to be part of a compromise between rules and regulations. But the consent required and collected is insufficiently informed, with devices arranged to encourage consent as much as possible. Even though the information for informed consent is present, it is not explicit, complete or understandable when clicking in a context of urgency and constant encouragement. However, the GDPR outlines requirements: "relating to processing in a concise, transparent, intelligible and easily accessible form, using clear and plain language".

Relations with digital and reticular technologies require explainability, which is, according to the CNIL (the French Commission for Information Technology and Civil Liberties), "the ability to put together the elements taken into account by the AI system to produce a result and make them understandable" (see: www.cnil.fr/fr/definition/explicabilite-ia), to encourage self-determination of information[4]. This notion is put to valuable use in a 2014 report by the French State Council on "Digital Technologies and Fundamental Rights"[5] and defined as the capacity of an individual to make decisions about the use of their personal data, based on rights to information, access, correction and opposition.

However, beyond the timing of the appropriation and debate, digital technology users must hastily approve the

4 See: https://www.cnil.fr/fr/ce-que-change-la-loi-pour-une-republique-numerique-pour-la-protection-des-donnees-personnelles.

5 See: https://www.conseil-etat.fr/actualites/discours-et-interventions/le-numerique-et-les-droits-fondamentaux. Also see "l'autodétermination informationnelle à l'ère de l'Internet" dated November 18, 2004: https://rm.coe.int/16806ae51f.

General Conditions of Use, a step that is often considered irritating by users who accept the conditions of traceability of their use to, in exchange, quickly benefit from digital content or services, with of course the possibility of configuring or refusing them, remaining watchful or even self-protecting[6]. Laws and regulations set out the limits, and civil society expresses, albeit diffusely, the possibility of refusing to be constantly tracked, to protect data and privacy.

According to a 2021 CNIL report, the authority was "subject to particularly intense activity and increasing solicitation" (see: www.cnil.fr/fr/la-cnil-publie-son-rapport-dactivite-2021). In 2022, the CNIL's report highlighted the increase in complaints and their processing (see: www.cnil.fr/fr/le-rapport-annuel-2022-de-la-cnil). However, digital transformations with connected objects and artificial intelligence systems orchestrate pervasive networks – interconnected services offering contextualized data – which means that simply referring to regulations does not suffice. Moreover, one of the actors of Internet governance, AFNIC (the French Association for Cooperative Internet Naming), proposed the creation of a "Federated ONS – Object naming service" architecture, using the Domain Name System (DNS) infrastructure to associate login details with the Internet of Things (IoT), representing a solution to address "the challenges of security and confidentiality protection posed by connected objects"[7].

Tensions related to the intensification of the digitalization of society, now also in terms of datafication (the process of turning all social activities into data), are therefore running very high. At the same time, the development of applications and digital services is accompanied by user-centered

6 See VPN and "onion routing": https://www.cnil.fr/sites/cnil/files/2024-03/cnil_guide_securite_personnelle_ven_0.pdf.

7 See: https://www.afnic.fr/observatoire-ressources/actualites/le-dns-la-solution-de-nommage-de-linternet-des-objets-le-point-de-vue-de-lafnic/.

design to encourage so-called intuitive uses, but without informed co-decision, thereby distancing user participation. Notwithstanding this, the digital commons movement "fights for the reinforcement of citizens' power to act through sharing, equality of access, and self-government of digital devices":

> This movement calls into question the logical processes of the market and the State in the administration of certain shared resources and proposes to establish a third political paradigm (Shulz 2021).

This sociopolitical dynamic is acknowledged by the government[8], indicating that users have been included in the process of defining the rules of design and use to stimulate expression and creativity. Concurrently, these inclusion-based policies and strategies serve to extract user data, which is subject to the model of constant innovation, the latest of which concerns generative AI. These tensions and claims illustrate the ongoing nature of power relations in society and also relate to connected objects in sensitive sectors such as health. Indeed, the ideology that promotes belief in technological solutionism to improve the reliability of predictions and prevention – as calculation power is concentrated – opens the door to the datafication of activities in society, using networks and connected objects subjected to algorithmic processing and AI systems to derive advantages in terms of assistance with decision-making – in the case of health, for care.

In the health sector, for both health professionals and patient-users, connected objects are technical mediators of the doctor–patient relationship. Connected objects encourage willing participation in a digital society, with the power of

8 See: https://labo.societenumerique.gouv.fr/fr/articles/les-communs-num%C3%A9riques-un-mod%C3%A8le-innovant-de-d%C3%A9veloppement-des-ressources-num%C3%A9riques/.

quantification, disrupting the way policies are created for health in social life through flows of health data (monitored, recorded and collected for multiple interpretations and the legitimization of the process) in a regime of presence permanently controlled by robots. E-health relies on the foundation of the innovation market to instill belief in saving the health system, leading to consent from professionals and patients, who see in this a source of predictions and risk detection. The CNIL believes that national-scale digital projects in health will have "a significant impact on public and private actors and on society as a whole" (see: www.cnil.fr/fr/la-cnil-publie-son-rapport-dactivite-2021; www.cnil.fr/fr/le-rapport-annuel-2022-de-la-cnil).

Covid-19, which required adaptation of information systems and the "Accelerating the Digital Shift in Health" roadmap (see: www.esante.gouv.fr/virage-numerique/feuille-de-route), was a source of major challenges, considering the National Health Data System (SNDS) created by the Health System Modernization Law (no. 2016-41 of January 26, 2016). However, looking at interoperability in particular, the security of health data – and also medical and administrative – may be insufficient. Additionally, the CNIL recorded notifications, which more than doubled in 2021 (see: www.cnil.fr/fr/la-cnil-publie-son-rapport-dactivite-2021, p. 90), demonstrating that sensitive health data had been compromised.

The CNIL processes hundreds of authorization requests in the health sector while continuing its work overseeing the management of health files (digital patient file, shared medical file, entitled "My Health Account"). In the era of strengthened authentication processes, with the "pro Health connect" service developed by the French Digital Health Agency, the CNIL carries out its activities overseeing the security of health data in medical analysis laboratories, hospitals[9], and with their service providers, the health data brokers[10].

All these precautions have not prevented the CNIL from authorizing Microsoft to host the health data of French citizens (Health Data Hub) in February 2024[11], which provoked certain reactions, in particular from ISOC France (the French chapter created in 1996 by the Internet Society, the association representing Internet users in Internet governance bodies), which submitted a claim to the French State Council to have this authorization[12] revoked.

This increase in protection and support activities demonstrates the intensity of the highly sought-after health data market. The e-health market contains an increasing number of innovations in the field, such as connected objects in health used by doctors and patients (Carré et al. 2021), who know that their data are being tracked (in the medical context) and generally have confidence in the goodwill of doctors and the hospital with regard to their data (despite leaks and cyberattacks). They wish to be actors in their own healthcare, in accordance with a certain empowerment through digital technology.

The term "empowerment" used in the digital technology regime presents a "conceptual and semantic vagueness [...]

9 For example, relations with the Department of Clinical Research and Innovation, DRCI at AP-HP. See: https://recherche.aphp.fr/la-delegation-a-la-recherche-clinique-a-linnovation-de-lap-hp/. See also the CPP (committees for the protection of persons): https://cncpp.fr.

10 Monetization of tracked data. See: https://www.cnil.fr/fr/cookies-et-autres-traceurs/regles/cookie-walls/monetisation-des-donnees-personnelles-les-enjeux-juridiques-et-ethiques.

11 See: https://www.cnil.fr/fr/la-plateforme-des-donnees-de-sante-pds. After the request for guarantees by the French State Council, see: https://www.cnil.fr/fr/le-conseil-detat-demande-au-health-data-hub-des-garanties-supplementaires.

12 See: https://www.isoc.fr/wp-content/uploads/2024/02/CP_ISOCFR_Donneesdesante.pdf.

an ideological shifting of the notion towards more liberal meanings [...] which simultaneously weaken its critical and radical influence" (Labarthe and Trainoir 2019, paragraph 6). Doctors, for whom human intervention is of primary importance in medical diagnosis (Lindenmeyer 2022), look to connected devices as a way to save time, be more responsive and avoid certain appointments, but at the same time these devices lead to cognitive saturation. They acknowledge that they do not have total control over the technological environment, in particular when they envisage the contribution of artificial intelligence (as a decision-making aid, automation of diagnoses, merging of mass data).

Doctors are also aware of data protection for their patients and of privacy by design. But the storage of health data (other than for scientific research or reasons of public interest) may target the construction, operation and continuous improvement of machine learning algorithms. In the latter case, compliance formalities must be completed or authorization by the CNIL is required, with the consent of those concerned, even though the data may be anonymized or pseudonymized, in order to limit the risk of data reidentification.

With AI, there is a risk that informed consent is limited, insofar as all the elements in connected devices are not (currently) fully understood, making it difficult to guarantee data processing compliance. In 2021, the CNIL, which also provides assistance to the insurance sector given the challenges associated with health data processing by these actors, and which chairs "compliance clubs" with the main representatives in the sector (French insurers, the French Technical Center of Contingency Insurance Institutions, the National Federation of French Mutual Insurers), created a "sandbox" of personal health data in order to support projects in the field of digital health in terms of legal security and data protection, considering the innovative nature of the project, the benefit to the public and the benefit for data protection (see: www.cnil.fr/fr/bac-sable-donnees-personnelles-la-cnil-accompagne-12-projets-dans-le-domaine-de-la-sante-numerique).

Supporting the innovative digital health market and guaranteeing the protection of health data and privacy constitute a “compromise”, according to the CNIL, one of the regulatory authorities (see: www.economie.gouv.fr/cge/regulation-numerique) in France, which alone cannot avoid the risks to freedoms and rights. The intensifying deployment of AI challenges the main principles of the French Data Protection Law, the GDPR and European regulations on AI[13], which are under pressure.

The independent administrative authority points out practices prohibited by AI systems which have a high potential for manipulating people through social rating and the use of remote biometric identification systems “in real time” in areas accessible to the public, creating a high risk to health, safety and the fundamental rights of individuals[14]. Somewhere between fear and frenetic action, decision-makers and users seize on technologies that can do remarkable things – particularly through calculation speed – and represent an imagined (Flichy 2001) world that leads to an exhausting transformation of society under the control of technology (Ellul 1988), with considerable costs, particularly in terms of energy.

The ambivalence of information technology, from advantages to restrictions, results in a system in which the role of the technician (Ellul 2004) dictates behavior and conditions all activities in society. Thus, the ingenious

13 European Parliament (2022). AI and health. 21–36 [Online]. Available at europarl.europa.eu/doceo/document/TA-9-2022-0140_EN.html.

14 See: www.legifrance.gouv.fr/jorf/id/JORFTEXT000045593731, paragraph 16.

production of information technology brings about consent – not necessarily informed – and fascination can become an alienation, albeit partial (Lefèbvre 1961). But if awareness of partial alienation disappears, the alienation risks becoming complete (Lefèbvre 1980), through exaltation and inevitability, leading to thought channeled through the prism of information technology.

In the early 21st century, technological opacity is caught up in the swell of rules and regulations, coordination, planning, expertise and evaluations to attempt to frame powers and counterpowers in terms of both information technology and digital processing. Including in the health sector, connected objects and artificial intelligence systems, which are expected to save time, predict and provide decisions, involve delegation to machines in order to fulfill the utopia of avoiding human error, reducing uncertainty and entropy, recalling the cybernetics of Norbert Wiener (Breton 1996, p. 131). Computers and all connected objects are included in a system that must constantly be balanced (homeostasis) by feedback mechanisms for the purpose of adaptation (machine learning) and to target unknown or partially identified objectives (deep learning).

For users, it is a case of benefiting from services through machines, and for companies a case of benefiting from users' data, and orienting behaviors calculated for the prediction market (Zuboff 2020) – in summary, directing behaviors (Foucault 1982). Digital transformation with the Internet of Things thus concerns society as a whole: "control-based societies which are in the process of replacing discipline-based societies" (Deleuze 2003b, 2018), aiming for efficiency but posing risks to fundamental liberties.

The difference between traceability (with informed consent – such as the use of cookies, see: https://www.cnil.fr/sites/cnil/files/2024-03/cnil_guide_securite_personnelle_ven_0.pdf) and surveillance, then a shift from surveillance (sometimes at the limit of liberty-suppressing practices) to control – an implementation of power that serves different interests such as security and authorization, for both private and public actors – sheds light on the deployment of pervasive and automatic devices for data collection about populations, facilitating the creation of large volumes of data (big data) to extract maximum value from their sophisticated algorithm processing in terms of policy, economy and health.

Referring to the works of Foucault, Agamben understands the concept of the device (*dispositif*) as "a set of practices, knowledge, measures, and institutions that aim to manage, govern, control, and direct the behaviors, gestures, and thoughts of humans in a way that is intended to be useful" (Agamben 2006, paragraph 23).

The data market – including that of health – with its pervasive devices, joins the profiling market in the era of the so-called empowerment of Internet users and visible, consenting users, whereas monitoring strategies are opaque, draining away the promise of personalized online care or services through the automation and interconnection of the objects.

The normalization of digital monitoring and the fatigue of having to constantly read and configure general conditions of use and tick the box in the everyday rush dilute vigilance as a counterpower, which is handed back to regulators and legislators to protect rights, reflecting a belief in the bodies capable of protecting data and freedoms.

In return, rewards or incentives, with user satisfaction taking up an important position, allow activities to continue in connected environments and aim to improve service to those who have ticked their "informed consent" (in other words, the users know that they are tracked), accepting the conditions of surrendering privacy for personalized care, content and communication by digital networks.

For actors in the digital economy, monitoring is effective at all times and in all locations to provide security and serve connected users (health professionals and patients), who are both the object and the subject, producing themselves the means and the condition for this monitoring, within "control societies, which do not operate by becoming closed off, but by continuous inspection and instantaneous communication" (Deleuze 2003a, p. 236).

In a context of generalized acceleration, technology – as a component of human civilization – aims for efficiency (Ellul 1990, pp. 121–135). Indeed, automation and computing power, information technology networks and contemporary artificial intelligence systems form the environment that disrupts rhythms, ways of acting and reflexivity, prompting us to consider ourselves users and actors within this ecosystem, caught between attractiveness and fear.

Prescriptions are integrated into traceability and profiling devices ("prescription by design", "prescription markets" (Benghozi and Paris 2007, p. 12)), whether individuals are aware of it or not, and the uses of which are negotiated, often unequally, within pervasive environments: "The barometer of digital technology points to the idea that executives and intellectual professions are slightly more cautious in their effective practices, which can lead to a greater recourse to rights by these social groups" (see: www.cnil.fr/fr/la-cnil-publie-son-rapport-dactivite-2021, p. 60).

With digital technology networks, the illusion of autonomy created by connected objects reactivates the question of prescriptions, insofar as the individual must accept giving up some of their freedoms if they wish to acquire new ones (including the use of connected objects).

In this digital ecosystem, the health sector presents a considerable challenge in terms of protecting sensitive data, which is in constant circulation through connected objects related to health or well-being, and in the face of persistent confusion given that connected objects in health (Carré et al. 2021) must meet the provisions of French public health laws, whereas those falling within the well-being market are not subject to them.

Vigilance also involves reviving the DMP[15] (personal medical file; called "dossier patient informatisé", or DPI (see: www.cnil.fr/fr/donnees-de-sante-la-cnil-rappelle-les-mesures-de-securite-et-de-confidentialite-pour-lacces-au)) with "My Health Account." This case marks the ability of the State to reintroduce a system that was widely rejected by those primarily concerned. This new "health account" would be "beneficial" for citizens, assigned to them by default (opt-in), except if each citizen explicitly declares that they do not wish to have it (opt-out).

Patients have begun to appropriate this type of online "space", where secure health data sharing takes place[16], while maintaining their digital self-measurement practices (using devices and applications). The French High Authority for Health (HAS), created by the law of August 13, 2004, regarding health insurance, reports: "More than eight in ten people (are) equipped with a smartphone, 61% with a computer, 56% a tablet computer, 23% a connected object related to health" (see: www.linc.cnil.fr/fr/barometre-du-numerique-2021-les-chiffres-des-usages-numeriques-en-France).

15 Schweitzer, L. (2012). Le DMP ou comment constituer un gigantesque fichier des données de santé. *Terminal*, 111 [Online]. Available at: http://journals.openedition.org/terminal/1023.

16 See: https://www.monespacesante.fr/protection-donnees-personnelles?gclid=EAIaIQobChMIl9276fqG-QIV15TVCh2QXwM6EAAYASAEEgKnaPD_BwE&gclsrc=aw.ds. Not forgetting the CNIL's authorization in February 2024 for Microsoft to host the health data of French citizens. See: https://www.cnil.fr/fr/les-principaux-avis-et-recommandations-de-la-cnil-sur-la-plateforme-des-donnees-de-sante.

Digital innovation in the health sector presents itself as an economic solution that improves the care system through telemedicine (remote consultations and remote expertise) and e-health, defined by the HAS as "a vast domain of applications of information and communication technologies serving health... devices such as prescription assistance software and computerized medical files" (see: www.has-sante.fr/jcms/p_3460778/en/e-sante).

The HAS is, of course, aware of the risks to health data security; therefore, health data hosts are required to be approved. It is also necessary to establish reliable and robust information systems for public hospitals which, however, do not have all the financial and technical means required for protection and security.

In the health sector, care can be monitored and behavior controlled through anticipation and decision-support tools. The State has always monitored and implemented instruments to track the population (Vitalis and Mattelart 2014). Cybercontrol is thus at the service of algorithmic governmentality (Rouvroy and Berns 2013), representing algorithms that allow profiles to be created in order to classify and plan the heterogeneous behaviors of individuals, in favor of predictive medicine which is prepared to adapt care, including remotely.

In a context of hyperconnectivity (Carré and Vidal 2018), the trade-off between continuous monitoring and care is easy, but the unequal relations and games of domination remain unchanged with the opacity of programming[17]. There is an illusion of reciprocity, even though the renunciation of certain freedoms lies at the heart of digital uses, which are

17 With "algorithmic enclosures" and third- and fourth-generation algorithms [...] that are more complex and more subtly perverse [...] and more difficult to locate and identify" (Ertzscheid 2015, paragraphs 27 and 37).

negotiated to a diminishing extent (Vidal 2010, 2017). The feeling of being autonomous through the use of information technologies and digital networks outweighs the fear of being monitored, especially when assurances are given about the security of exchanged data in everyday life.

In accordance with the interactionist approach, Howard Becker suggests that the fabrication of social aspects draws on daily activity: "The activities of a person are often restricted by the type of role they have played earlier in the interaction. They believe they have staked their appearance on being a responsible and consistent participant, thus committing to an activity trajectory that is congruent with this role" (Becker 2006). Restriction and control are internalized by individuals, referring back to the rules incorporated as *habitus* (Bourdieu 1980), a socially constituted scheme that is decisive in the probable action of social subjects – here, being an actor in one's health by following the instructions of connected health objects.

Trajectories then take form and nevertheless nourish thought, attaching themselves particularly to an almost permanently connected daily life, without straying into the realm of alienation. According to the sociology of everyday nature developed by Henri Lefèbvre, "technological alienation" (Lefèbvre 1961, p. 211) contributes to reproducing the ideology of technological and economic progress, through which strategies ensuring the acceptance of networked digital technologies disseminated in society can be grasped. Humans thus adapt to be in agreement with machines – an assimilation that seeks to avoid "the aggression of technology" (Ellul 1990, p. 300).

Data processing using algorithms[18] and recent artificial intelligence systems, currently devoid of legal personality and

18 See the summary of the public debate of the CNIL regarding the ethical challenges of algorithms and artificial intelligence, held on December 15,

therefore unable to be held responsible for their actions, has nevertheless given rise to debates:

> Does artificial intelligence justify attributing legal personality to an autonomous robot that, however, does not have all the legal characteristics of an asset? Although the European Parliament responded positively by adopting the report of Deputy Mady Delvaux on February 16 of this year, the assertion of the legal nature of an asset for a robot constitutes a civilizational issue that must be defended[19].

In the context of data mining (for deducing and identifying trends and profiling) and machine learning (supervised or automated), communication networks in the health sector, composed of and between all types of (connected) objects, circulate a considerable amount of data arising from relations between humans or with non-human devices, from social activities to private and personal lives, forming a data market characterized by an accelerating movement absorbed by human societies.

6.2. Acceptability, ambivalence of uses

In this context of intensive computerization, the sociopolitical approach to digital uses highlights the political dimension of technical, economic and social logic of a computerized society. Viewpoints on connected objects and digital networks are maintained to persuade others that these technologies are justified and to self-proclaim their efficiency, exercising a performative force (Austin 1991). Without denying creative variants and alternatives, uses are conditioned by the

---

2017: https://www.cnil.fr/fr/comment-permettre-lhomme-de-garder-la-main-rapport-sur-les-enjeux-ethiques-des-algorithmes-et-de.

19 See: www.actu-juridique.fr/divers/le-robot-bien-ou-personne-un-enjeu-de-civilisation/.

organization of digital environments, which objectify individuals' capacity for information and communication – their relations through computerized networks, mediations and social activities, including those related to health.

Connected objects, developed even before their consequences, have been assessed, are attractive while presenting the (belief in) advantage, both for the decision-maker and the user, of social control, which, according to Howard Becker (2006) "consists of individuals obeying social rules [...] by postulating an interiorization of a generalized other that constitutes hidden assistance inciting compliance with the rules" (Becker 2006, paragraph 11).

This is how the acceptability of connected objects operates, including those designed with programmed obsolescence, which leads to rapid, abundant and incessant changes. With connected health objects, the communication mediated by interconnected services transforms patient-users into receivers and opinion creators, replacing the exchange during the follow-up visit with the doctor. These communication technologies, promoted as social innovations, claim to recognize an active audience supported by the values of performance and speed, key issues in an era of universalization of instrumental reason and of "acting in compliance with regulations" (Habermas 1973, p. 23).

Social acceptability can also, in a certain way, align with the double bind described by the Palo Alto school of thought (Winkin 2008), when individuals are ordered to experience an undesired situation – to be tracked without wanting to be tracked, to protect their data and privacy and simultaneously taking risks with them. This paradoxical injunction results in an inability to refuse remote medical monitoring by care devices.

Yet, emitters and receivers do not have the same equipment, the same command of software and functionalities, or the same knowledge of the algorithms that process their data. Of course, some expert users manage to identify the environment that records their data and thus ask questions of doctors or technical service providers – or even choose the connected object in the health system that monitors their condition. Nevertheless, other patient-users of connected objects say they feel like actors in their own health[20], although their actions are limited to the commands recognized by the system, to personalized content or to the self-configuration of the connected care device.

This last point embodies the domination that has guided its design, creation and production – in technological and economic opacity. Uses are shaped by information technology environments and by the storage of decision-oriented and profiling-oriented data for personalized and intrusive medical monitoring, accepted, on the one hand, by doctors who adopt them in their practice and management of patients, and, on the other hand, by patients who have provided signed consent.

Socially, maximum value is drawn from the connected objects in their lives; but if the exchange is disrupted, the user must then find ways to deal with new codes and non-intuitive methods. If these constraints are rejected or criticized, users may sometimes engage in a creative relationship with the device through tactics intended to avoid, bypass, resolve or draw benefit from the problems that arise (e.g. discovering a function not accessible via the main interface, such as printing results to view them outside the connected device). On the other hand, through such rejection, the patient risks being unable to benefit from regular remote monitoring.

20 See the WHO Europe website: https://www.who.int/europe/en/initiatives/empowerment-through-digital-health.

Without doctor–patient appointments, through "transactions" with technical devices, users experience the immediacy of access with a sense of meaning for their professional (doctors) or personal (patients) lives. On the other hand, patients who are unable to use their connected health devices feel very frustrated. Indeed, while waiting for a medical visit or technical support, those who do not give up embark on a quest to master the technology, discovering settings and other functions.

This engagement in technical mediation allows them to read content from the connected object without engaging in dialogue with doctors. Certain patient-users sometimes express themselves on platforms (digital social networks) that allow new relationships to be built with information and content, thanks to their online participation and contribution skills. Users are thus invited to take action, guided by the imaginary world of technical potential, which plays a driving role in the desire to access content that is difficult to reach, or new, or to become partners of the connection of objects that link the private or personal sphere with the outside world – in the case of connected health objects, the medical sphere.

This imaginary plays a role in the process of appropriating connected objects, and a certain narcissism combines with a feeling of efficiency thanks to interfaces where content can be manipulated (supposedly) exactly as desired. When the user, whether a professional or not, adopts a critical attitude, they demonstrate that they are not captivated by the technology, even if they claim to be empowered through the connected object, which can become a constraint or a burden, in other words, something endured.

Users, motivated by the representation of digital technology as a "territory" of possibilities, adopt their connected health object as part of their information technology and digital practices, as they have done with other digital industrial consumer products. Little by little, users' "microscopic" actions, to use De Certeau's (1990) term, take their place in a slow process of transformation in information and communication methods, particularly concerning their health and self-treatment.

Engagement in digital mediation, during which the body[21] positioned in front of the computer comprises one of its instruments, confers on users the status of actors – producers of meaning through their appointments and uses – anchored in advance in frameworks of knowledge and Internet practices or in practices related to other connected well-being devices (through interactions of practices, knowledge of the health system, and relationships with doctors and hospitals).

Screens appear as interfaces offering a variety of content, and connected health objects are part of a set of other connected devices from various fields: cultural, educational, commercial, recreational, administrative or professional. Thus, with almost nothing left to hide, Internet users and connected-object users participate in maintaining the digital ecosystem, which adapts and adjusts itself through perpetual change while setting the pace for social relations.

By trivializing the circulation of personal and sensitive data in this way, there is a devaluation or instrumentalization of health data that excludes the doctor–patient relationship. Unable to fully master human–machine dialogue, both professional and patient users decide to abandon explainability[22] – an understanding of how designers have planned uses and the possibility of accessing all content.

21 CNIL (2014). Le corps, nouvel objet connecté. *Cahiers IP*, 2 [Online]. Available at: https://www.cnil.fr/sites/cnil/files/typo/document/CNIL_CAHIERS_IP2_WEB.pdf.
22 See: https://www.polytechnique-insights.com/tribunes/digital/quels-enjeux-quand-les-algorithmes-remplacent-lhumain/.

It is not necessary for this renunciation to be regretted by users, who see it as a way to free themselves from many constraints (multiple appointments, hospital visits) and as a decision allowing them to adopt a comfortable posture as both order givers and content receivers. At this point, users, whether professionals or not, are both hostages and protagonists in the development of the economic value of all social activities, including healthcare.

Connected technologies undoubtedly save doctors time, as they delegate part of their relationships with patients to service providers who manage connected machines "capable of manufacturing consent" (Chomsky 2004, p. 33). The circulation of content and the enjoyment of easy access to health data brought about by user actions are, however, very real. The promise to deliver content is fulfilled, but the relationship is not necessarily established, as it is invoked by the device users in a compromise between liberties and constraints.

Injunctions such as human–machine dialogue, design-related constraints and usage rules can constitute a simulacrum in which the pleasure of immediacy (hoped for, sometimes achieved) competes with the acceptability and appropriation of technologies, within a volatile relationship in the context of communicational practices.

The situation is thus ambivalent, insofar as individual liberties of choice regarding appointments, access to information and expression are broadened and at the same time provide the basis for social control – associated with a form of biopolitics (Foucault 1982) that aims for the good health of protagonists in the digital ecosystem.

This ambivalence is part of a dual process of dominance and emancipation within the context of a normative power exerted through the subordination of individuals who are voluntarily connected and who collaborate with their own monitoring to benefit from digital services. The acceptability of monitoring – both endured and chosen (Carré and Panico 2012) – thus offers an opportunity for the full expansion of control without visible constraints, yet without eradicating criticism of the powerful actors in the digital economy[23].

The desire for digital objects and services leads, in effect, to "enthusiastic or addictive participation in and use of interconnected devices, applications, and internet platforms [...] always constrained and disciplined by the imperative of production and value maximization [...] [targeting] firstly and above all [...] the functions penetrated by market-based logic" (Richert 2019, p. 38).

Algorithmic calculations, connected objects and artificial intelligence constitute legitimate contemporary contexts for domination. In a generalized belief in networked technologies, with the awareness to protect personal data and privacy, users are driven towards willing participation and consent – sources of profit in economic terms for some, and of risk prevention and medical monitoring for others – in a context of abundance and acceleration through machines that have the capacity to self-regulate what can be predefined (connected health objects in accordance with care) and to fragment.

The world market for "information technology devices [...] tends to homogenize the world and simultaneously to fragment it" (Lefèbvre 1981, p. 145). This fragmentation at the microsocial level, involving connected health objects,

---

23 Concerning GAFAM, see the proposed dismantling at: https://www.vie-publique.fr/parole-dexpert/281851-les-gafam-vers-une-regulation-ou-un-demantelement-par-louis-perez. Also, regarding BATHX: see the 2019 information report on the "giants of digital technology", presented to the *Assemblée nationale* (French lower house of parliament) by Members of Parliament Alain David and Marion Lenne: https://www.assemblee-nationale.fr/dyn/15/rapports/cion_afetr/l15b4213_rapport-information#_Toc256000021.

combines with fragmentation at the macrosocial level, concerning public space and social relations; and at the meso-level, with the strategies of economic actors supported by political actors favorable to the daily expansion of these technologies, through "the integration of uses in social relations" (Jouët 2000, p. 499).

By studying uses – particularly the contributions of patients equipped with connected health objects – we can move beyond the question of access, which is most often treated quantitatively when evaluating their roll-out across society, and instead promote the social acceptability of these connected machines in order to legitimize them.

The ambivalence of the uses of connected objects in healthcare involves both volatility and stability, a taste for innovation and a need for benchmarks and guaranteed, secure data (since sensitive health data are subject to the highest level of privacy protection), a desire for acceleration, and the simultaneous acceptance and rejection of prescription. Political and economic decision-makers who deploy programs, roadmaps and viewpoints on digital technology to support the health sector win over the loyalty of the population while justifying the limitations on human liberties required to achieve their objectives in terms of public health and the e-health market.

The technical and economic questions raised by experts take precedence over public debate, to the point of eliminating it entirely. This neither prevents nor provokes questions: institutional communications, online consultations and digital support services are now available[24].

24 See: https://esante.gouv.fr/strategie-nationale, https://esante.gouv.fr/strategie-nationale/ethique-et-numerique, https://esante.gouv.fr/sites/default/files/media_entity/documents/dns-feuille-de-route-2023-2027.pdf, https://esante.gouv.fr/sites/default/files/media_entity/documents/210713_cp_ans_vf_0.pdf, https://www.cnil.fr/sites/cnil/files/202307/bilan_bac_a_sable_sante_numerique.pdf.

6.3. Sociopolitical challenges of the relationship between healthcare and digital technology: conditional uses and the capacity to act

Critical analysis of the uses of digital devices, including connected objects, takes into account the socioeconomic and political discourse and contexts of the computerization of society, up to the point where a computerized society is constructed – without assuming these forces to be deterministic, despite their determining and structuring effects. Contexts of use are subject to conditions, though these do not eliminate the capacity to act.

An example from the transport sector is enlightening: "In compliance with the regulations in effect, you may oppose this transfer. In this case, Île-de-France Mobilités will no longer be in a position to manage your contracts, and you will no longer be able to benefit from the services you currently enjoy (Navigo or Imagine R passes and the associated transport documents) for your travel" (see: www.iledefrance-mobilites.fr/). The same holds true across all sectors, including healthcare, even if exceptions are made for specific processing related to public health, national education, culture, freedom and the plurality of the press.

Thus, the beliefs of politicians and health professionals in technologies that save time for the e-health economy – or improve how they perform their caregiving duties – involves patients themselves in a dual delegation: first, to machines, and second, to remote medical care programs. Indeed, uses of connected healthcare objects form part of a matrix of viewpoints and actions that reproduce existing relationships and roles between doctors and patients. New behaviors within doctor–patient relationships, mediated by connected objects as interaction devices, aim for shared production of knowledge at the core of their use.

To this end, the device's protagonists legitimize the sociocognitive context of physically connected, material objects and integrate them into the daily life of both parties. However, as a disruptive innovation, connected healthcare objects unsettle cognitive benchmarks and established behaviors. It is therefore necessary to adapt the practices of each party by relying on reciprocal engagement.

From the perspective of the sociology of innovation, which departs from Foucault's disciplinary framework, the "scripts" (Akrich 1987) embedded in objects, along with their mediation and translation (Latour et al. 2006), infiltrate a sociotechnical network (the network actor), given the heterogeneity of the social organization that requires devices in order to act, adapting them as needed. Information technology infrastructures, systems and digital devices (applications and communication interfaces) "organize" the activities of individuals while still allowing them the possibility to take action (Beuscart and Peerbaye 2006, paragraph 3), thereby feeding the device or technical system (as in Ellul's analysis).

From this perspective, the appropriation of digital networks – particularly the Internet – can lead to a space for action where sociotechnical innovations unfold (such as peer-to-peer systems, blockchain or TOR (see: www.torproject.org/fr/)):

> The capacity to act – a concise way of defining agency – was reinforced by the availability of new technical methods which, although not originally intended for these purposes, have been adapted to serve them. [...] Activity in the public arena can now be extended to include technical questions previously considered neutral and left to the discretion of experts without consultation. This creates a social and technical environment in which agency in traditional political fields has also begun to liberate itself from the

passiveness induced by decades of a regular regime of radio propagation (Feenberg 2014, paragraph 39).

Concerning connected health objects, politicians and health professionals place their trust in technologies capable of predicting care actions, while patients trust in the doctors who monitor them remotely. The stakeholders in these digital systems accept the ongoing transformation of the healthcare sector and the fabrication of predictive models that evaluate effectiveness through the lens of interconnected machines. This belief is grounded in a pre-existing network of legitimate authority within healthcare – namely, doctors – as well as in a system vested with the power and legitimacy to make collective decisions on public health policies through debate, compromise and negotiation.

These developments are also supported by financial and technical backing; for example, the CNIL has strengthened its actions in line with a contemporary form of normativity[25], namely "dialogued normativity"[26]. All this is sustained through interlinked engagement – albeit unequal and competitive – among the actors behind these increasingly popular innovative technologies.

At this cost, both uncertainty (predictions) and the unknown (innovations) are accepted in order to enable remote interactions via connected objects managed by technical service providers, who implement medical, technical and usage-related prescriptions. This constitutes a "market of prescriptions" (Hatchuel 1995, pp. 12–15), which is socially viable because it is based on rules, legitimacy and power. These interactions sustain the socially accepted device through a "circular relationship between production

25 See: https://www.conseil-constitutionnel.fr/nouveaux-cahiers-du-conseil-constitutionnel/la-normativite-de-la-loi-une-exigence-democratique.

26 See: https://www.conseil-constitutionnel.fr/nouveaux-cahiers-du-conseil-constitutionnel/normativite-et-regulation.

and reception" (Ross 2013, paragraph 29). Service providers and doctors highlight the active role played by the public, who are represented in their absence, while patients, in turn, construct a representation of medical authority that legitimizes interaction through interposed connected objects.

However, acceleration within the technical system reduces action to a negotiation of uses that forms part of a programmed renunciation, one that requires consent while relying on GDPR regulations intended to provide protection. The concept of negotiated renunciation (Vidal 2010, 2017) reflects the ambivalence inherent in information technology and digital practices – otherwise referred to as the social acceptability of the datafication of all social activities – while fostering a reflective stance towards the use of regulated monitoring.

From the notion of a social contract designed to liberate individuals to the control exerted over them by algorithmic governmentality, which constrains them, social alternatives emerge, such as informational self-determination – based on a model of social autonomation that renews the protection of personal data (through awareness and decision-making regarding its processing). In the healthcare context, the use of data is essential and constitutes an integral part of the monitoring mechanism – in this case, a medical one.

Regulatory limits are bulwarks against the growing intensity and complexity of society's datafication processes. While information about the rights to protect personal and sensitive data, as well as about the ethics of information technology and AI, is widely circulated, there remains a risk of resignation in the face of hyperconnectivity – seen by many as inevitable – and the proliferation of regulatory contexts and frameworks, along with the generalized monitoring and processing of all data, which will only continue to expand.

To ensure this type of protection, cybersecurity must be a central concern in combating cyberattacks. In France, the website cybermalveillance.gouv.fr, working in collaboration with the CNIL, aims to guarantee IT security. At a supranational scale, the Network and Information System Security (NIS) directive, adopted in July 2016, revised in October 2021, and transposed into French law in 2018, pursues the same objectives within the European Union[27].

We should also note the role of regulatory and mediation bodies such as the European Data Production Supervisor (EDPS), data protection officers (DPOs, formerly known as data protection correspondents or CILs) and the French National Committee for Digital Ethics (CNPEN). Concerning AI, the global partnership on AI (GPAI.ai) also plays a key role, among other ongoing regulatory initiatives. Finally, media and digital literacy[28] remain essential to strengthen the vigilance and critical awareness of all users of connected objects.

Connected objects in healthcare serve a noble purpose, yet they form part of a sociotechnical system that encompasses material infrastructures, functional interfaces, conditions of use, and the accompanying rules and laws – all aimed at fostering political and economic action conducive to new forms of governmentality. This system reassures and monitors within an organized network of human actors (disciplined doctors and patients, whose actions are continuously guided and observed, even as they remain

27 See: https://www.ssi.gouv.fr/entreprise/reglementation/directive-nis/, https://www.ssi.gouv.fr/actualite/revision-de-la-directive-nis-une-opportunite-pour-renforcer-le-niveau-de-cybersecurite-au-sein-de-lue/, https://www.cnil.fr/fr/notifications-dincidents-de-securite-aux-autorites-de-regulation-comment-sorganiser-et-qui-sadresser, https://eur-lex.europa.eu/legal-content/FR/TXT/?uri=CELEX%3A32016L1148.

28 See: https://www.arcom.fr/se-documenter/ressources-pedagogiques/education-et-citoyennete-numerique/kit-pedagogique-du-citoyen-numerique-retrouvez-toutes-les-ressources.

participants in the process) and of machines, operating between the determined and the indeterminate.

The strategic urgency surrounding connected healthcare objects may, however, give rise to situations that escape doctors' original intentions. These technologies emerge from collective actions – both explicit and implicit, visible and invisible – drawing on knowledge and data to shape relationships between protagonists, who, in turn, acquire new reflexive capacities and transform their interactions. Contributions and constraints alike are subject to constant adjustment, involving external actors whose interventions introduce fragility into the once unique doctor–patient relationship.

The actions of various endogenous and exogenous stakeholders within this system, including those of actors in the e-health market, pose threats to individual liberties and privacy, generating risks that extend beyond existing regulatory frameworks. Interdependencies set new chains of responsibilities in motion across technical, medical, institutional and economic domains, while also activating the patient's own responsibility. The engineering of connected healthcare objects thus depends on the interplay between visible and invisible actors to construct a strategic apparatus that none fully comprehends. What endures is a persistent belief in sociotechnical solutions as the path towards implementing 21st-century healthcare policy.

6.4. References

Agamben, G. (2006). Théorie des dispositifs. *Po&sie*, 1(115), 25–33 [Online]. Available at: https://www.cairn.info/revue-poesie-2006-1-page-25.htm.

Akrich, M. (1987). Comment décrire les objets techniques ? *Techniques et culture*, 9, 49–64 [Online]. Available at: https://halshs.archives-ouvertes.fr/halshs-00005830/document.

Austin, J. (1991). *Quand dire c'est faire*. Le Seuil, Paris.

Becker, H. (2006). Notes sur le concept d'engagement. *Tracés*, 11 [Online]. Available at: http://journals.openedition.org/traces/257 (Notes on the concept of commitment (1960). *American Journal of Sociology*, 66, 32–40. Sur le concept d'engagement (2006). *SociologieS*, Redécouvertes [Online]. Available at: http://journals.openedition.org/sociologies/642).

Benghozi, P.-J. and Paris, T. (2007). The economics and business models of prescription in the Internet. In *Internet and Digital Economics: Principles, Methods and Applications*, Brousseau, E. and Curien, N. (eds). Cambridge University Press, Cambridge, 291–310 [Online]. Available at: https://hal.archives-ouvertes.fr/hal-00263198.

Benghozi, P.-J., Bureau, S., Massit-Folléa, F. (2012). *L'Internet des objets*. Éditions de la Maison des sciences de l'homme, Paris [Online]. Available at: https://books.openedition.org/editionsmsh/78.

Beuscart, J.-S. and Peerbaye, A. (2006). Histoires de dispositifs (introduction). *Terrains & travaux*, 2(11), 3–15 [Online]. Available at: https://www.cairn.info/revue-terrains-et-travaux-2006-2-page-3.htm.

Bévière-Boyer, B. (2021). La protection des données de santé mises à disposition par le Health Data Hub pour les recherches sur la Covid-19. *Journal du Droit de la Santé et de l'Assurance Maladie* (JDSAM), 2021/2(29), 37–48 [Online]. Available at: https://www.cairn.info/revue-journal-du-droit-de-la-sante-et-de-l-assurance-maladie-2021-2-page-37.htm.

Bourdieu, P. (1980). *Le sens pratique*. Éditions de Minuit, Paris.

Breton, P. (1996). L'utopie de la communication entre l'idéal de la fusion et la recherche de la transparence. *Quaderni*, 28, 125–133.

Carré, D. and Lacroix, J.-G. (1999). Virage ambulatoire et autoroutes de l'information dans le secteur de la santé. *Sciences de la société*, 47, 23–41.

Carré, D. and Panico, R. (2012). Du fichage subi, à l'affichage de soi. Eléments pour une approche communicationnelle du contrôle social. In *Connexions. Communication numérique et lien social*, Proulx, S. and Klein, A. (eds). Presses universitaires de Namur, Namur, 269–283.

Carré, D. and Vidal, G. (2018). *Hyperconnectivity: Economical, Social and Environmental Challenges*. ISTE Ltd, London, and John Wiley & Sons, New York.

Carré, D., Vidal, G., Sandré, S. (2021). Recherche sur les objets connectés en santé (ROCS-SHS). Research Report, AP-HP, Paris.

de Certeau, M. (1990). *L'invention du quotidien*. Gallimard, Paris.

Chomsky, N. (2004). Les exploits de la propagande. In *Propagande, médias et démocratie*, Chomsky, N. and McChesney, W.R. (eds). Écosociété, Montreal, 13–67.

CNIL (1978). Rapport de la CNIL, bilan et perspectives (1978-1980) [Online]. Available at: https://www.cnil.fr/sites/cnil/files/atoms/files/20171116_rapport_annuel_cnil_-_1er_rapport_dactivite_1978-1980_vd.pdf.

Deleuze, G. (2003a). *Pourparlers*. Éditions de Minuit, Paris.

Deleuze, G. (2003b). Post-scriptum sur les sociétés de contrôle. In *Pourparlers*. Éditions de Minuit, Paris, 5.

Deleuze, G. (2018). Les sociétés de contrôle. *EcoRev'*, 46(1), 5–12 [Online]. Available at: https://www.cairn.info/revue-ecorev-2018-1-page-5.htm.

Ellul, J. (1988). *Le bluff technologique*. Hachette, Paris.

Ellul, J. (1990). *La technique ou l'enjeu du siècle*. Economica, Paris.

Ellul, J. (2004). *Le système technicien*. Calmann-Lévy, Le Cherche Midi, Paris.

Ertzscheid, O. (2015). Usages de l'information numérique : comprendre les nouvelles enclosures algorithmiques pour mieux s'en libérer. *Revue française des sciences de l'information et de la communication*, 6 [Online]. Available at: http://journals.openedition.org/rfsic/1425.

Ertzscheid, O. (2017). *L'appétit des géants. Pouvoir des algorithmes, ambitions des plateformes*. C&F éditions, Blogollection, Caen.

Feenberg, A. (2014). Vers une théorie critique de l'Internet. *Tic & société*, 8(1–2) [Online]. Available at: http://journals.openedition.org/ticetsociete/1382.

Flichy, P. (2001). La place de l'imaginaire dans l'action technique. Le cas de l'Internet. *Réseaux*, 5(109), 52–73.

Foucault, M. (1982). Le sujet et le pouvoir. In *Dits et écrits, II*. Édition Quarto Gallimard, Paris, 1041–1062.

Habermas, J. (1973). *La technique et la science comme "idéologie"*. Gallimard, Paris.

Hatchuel, A. (1995). Les marchés à prescripteurs [Online]. Available at: https://minesparis-psl.hal.science/hal-01863931/file/Les%20marche%cc%81s%20a%cc%80%20prescripteurs%20(1).pdf.

Jouët, J. (2000). Retour critique sur la sociologie des usages. *Réseaux*, 18(100), 487–521.

Labarthe, F. and Trainoir, M. (2019). Explorer les déclinaisons de l'empowerment en régime numérique. *Terminal*, 125–126 [Online]. Available at: http://journals.openedition.org/terminal/5570.

Latour, B., Callon, M., Akrich, M. (2006). *Sociologie de la traduction*. Presses de l'École des Mines, Paris.

Lefèbvre H. (1961). *Critique de la vie quotidienne, tome 2. Fondements d'une sociologie de la quotidienneté*. L'Arche, Paris.

Lefèbvre, H. (1980). Aliénation et société. *Raison présente*, 55, 101–103 [Online]. Available at: https://www.persee.fr/doc/raipr_0033-9075_1980_num_55_1_2089.

Lefèbvre, H. (1981). *Critique de la vie quotidienne, tome 3. De la modernité au modernisme (Pour une métaphilosophie du quotidien)*. L'Arche, Paris.

Lindenmeyer, C. (ed.) (2022). Enjeux et effets des dispositifs numériques sur la relation médicale au prisme du traitement du syndrome des apnées du sommeil. Research Report [Online]. Available at: https://cis.cnrs.fr/wp-content/uploads/2022/06/vf-rapport-recherche-cis-ilpem.pdf.

Maise, H. and Vitalis, A. (1985). Les libertés : enjeu d'une société informatisée. *Études*, 4(362), 471–482 [Online]. Available at: http://gallica.bnf.fr/ark:/12148/bpt6k442013s.image.r=j%C3%A9sus.f41.pagination.langFR.

Mumford, L. (1963). Techniques autoritaires et démocratiques. Discours prononcé à New York le 21 janvier 1963 et publié dans la revue *Technique et Culture* (*Technology and Culture*), 5(1), 1–8, and in *Notes & Morceaux choisis*, 11, 26–24 [Online]. Available at: https://sniadecki.files.wordpress.com/2012/04/mumford_ums.pdf.

Richert, F. (2019). Big Data as a device for generalized decoding of the social field. In *Digitalization of Society and Socio-political Issues 1*, George, É. (ed.). ISTE Ltd, London, and John Wiley & Sons, New York.

Ross, P. (2013). Coordination entre production et réception. *Tic & société*, 7(1) [Online]. Available at: http://journals.openedition.org/ticetsociete/1282.

Rouvroy, A. and Berns, T. (2013). Gouvernementalité algorithmique et perspec d'émancipation. Le disparate comme condition d'individuation par la relation ? *Réseaux*, 1(177), 163–196 [Online]. Available at: https://www.cairn.info/revue-reseaux-2013-1-page-163.htm.

Shulz, S. (2021). Transformer l'État par les communs numériques : Socio d'un mouvement réformateur entre droit, technologie et politique (1990-2020). Sociology PhD Thesis, Université Gustave Eiffel, Champs-sur-Marne [Online]. Available at: https://www.theses.fr/2021UEFL2027.

Vidal, G. (2010). Le renoncement négocié. Pour une analyse dialectique des usages des technologies interactives. Information and Communication Sciences HDR Thesis, Université Bordeaux 3.

Vidal, G. (2017). Prendre la mesure du renoncement négocié. *Multitudes*, 68, 54–59 [Online]. Available at: https://www.cairn.info/revue-multitudes-2017-3-page-54.htm.

Vidal, G. and Panico, R. (2019). Digital practices, cultural practices, under surveillance. In *Digitalization of Society and Socio-political Issues 1*, George, É. (ed.). ISTE Ltd, London, and John Wiley & Sons, New York.

Vitalis, A. (2009). Informatique et libertés : une histoire de trente ans. *Hermès, la revue*, 53, 137–143 [Online]. Available at: https://www.cairn.info/revue-hermes-la-revue-2009-1-page-137.htm.

Vitalis, A. and Mattelart, A. (2014). *Le profilage des populations. Du livret ouvrier au cybercontrôle*. La Découverte, Paris.

Winkin, Y. (2008). La réception de la notion de double contrainte en France. In *La double contrainte. L'influence des paradoxes de Bateson en Sciences humaines*, Wittezaele, J.-J. (ed.). De Boeck Supérieur, Louvain-la-Neuve, 61–67 [Online]. Available at: https://www.cairn.info/la-double-contrainte--9782804157135-page-61.htm.

Zuboff, S. (2020). *L'âge du capitalisme de surveillance*. Éditions Zulma, Paris.

Conclusion

A Digital Prescription for All…

The undeniable advantages of contemporary technological developments – and their promises for the health sector, including rapid decision-making, precision medicine, pathology management on a population scale, improved regulation of pressure on health systems, and the indirect optimization of health insurance and related risk management – make digital connected devices both unavoidable and imperative. Their motive and justification are evident: they serve the health of the wider public.

The health field did not hesitate to dive in:

– relational and communicational technicization and rationalization: these result in a technical infrastructure which encourages automation of the selection processes inherent in medical practices, such as pathology and patient triage practices, processes followed to strengthen the technical infrastructure that has been implemented and processes that are not of interest or not taken into account because they do not have the advantage of supporting the expansion of this infrastructure. There is an ensuing reconfiguration of care practices, of the presence and distance in the relationship between doctors and patients, in a context of increasing

Conclusion written by Dominique CARRÉ, Geneviève VIDAL, Christian PAPILLOUD, Yanita ANDONOVA, Étienne HIEN and Sarah SANDRÉ.

integration of devices and objects connected to traditional medical equipment, as seen in the case of medical examinations and surgery, for example;

– datafication of all activity and automation of use of these data optimize the attribution tasks inherent in medical practice, whether this is the distribution of patients in various medical departments or monitoring of changes in illnesses by the various bodies in the healthcare system, including monitoring the spread of disease. Now largely supported by artificial intelligence (AI), medical practice not only manages to automate repetitive tasks such as establishing diagnoses, but it also diversifies in contact with data towards manufacturing indicators to predict and provide alerts for changes in the pathologies present and the emergence of new medical issues, as well as the remote treatment of patients, for example, in medical deserts or cases of isolated patients.

Faced with economic, social, organizational and political problems, the increasing use of connected devices and objects promises optimal implementation of technical feats backed by the definition of "noble" objects used for the purpose of generalized well-being. In this way, an image can be drawn of medicine rising to the challenge of its own modernization – of cutting-edge medicine that is able to improve the quality of care it provides, improve patient autonomy, enhance screening of pathologies and strengthen monitoring of chronic serious conditions (cancer, multiple sclerosis). At the same time, it encourages us to forget that the networked universe of this technical infrastructure involves challenges other than those specific to medicine alone – challenges that are also of another order. These are financial, commercial and industrial challenges that result from the dependence of health on digital industries – on companies that use the data from the technological environment in healthcare – without addressing the ethical and legal challenges that these technologies now raise on multiple levels. In terms of actors, such as doctors and patients, as well as social bodies in the health system, such as hospitals or clinics, they must increasingly dedicate some of their activities to managing technologies and data they host and share, as distinct from those they delegate to the private sector.

In different forms, the questions asked here evoke those raised at the time of implementing the "information highways" at the end of the 20th century, which saw the emergence of remote medicine. More than a quarter of a century after the "information highways", we can see that nothing has, nor will, slowed the expansion of connected devices and objects in the field of health, despite the risks associated with data protection and traceability of data processing for use by a health economy that extends beyond the medical arena into other fields, as we have come to understand. Of course, we can mention the problem of the social acceptability of connected devices and objects – of confidence and mistrust, of the resistance both for and against that they can provoke in society – a well-known issue in the field of humanities and social sciences. But, as we have recalled in this book, what can we really put in the way of an infrastructure whose expansion also represents a guarantee of collective well-being? It is not difficult to imagine that the indirect consequence of this increase in use will be progressive generalization of individual control – of both sick and healthy people – throughout their lifetime. Because the same is true here as in other technological fields: the lever for the effectiveness of all treatment, all medication, all therapy, is based on the quantity of information available. Only then will it be possible to better orientate decision-making, not only by doctors but also by insurance providers, pharmaceutical industries and digital companies, by exploiting the sociobiological profiles of social subjects and what they will teach us about various pathologies – their emergence, evolution or disappearance. Even by envisaging this development, which, as exaggerated as it may seem, nevertheless represents the *telos* of all connected devices and objects in our contemporary societies – Who would oppose this? Who would dare to risk cutting themselves off from the gesture or treatment that one day might save their life?

Health is priceless, and it is in this price that resignation is expressed – one that contemporary positive thinking reformulates as resilience, a word that nevertheless conveys the same reality: a collective renunciation of the social and cultural regulation of medical practice in favor of the medical regulation of connected devices and objects in health. In this context, there is nothing surprising in seeing more and more individuals equip themselves with smartwatches or seek information on the Internet and digital social networks about the pathologies they have – or might develop – in the hope of improving their understanding of what awaits them and their ability to discuss it with medical personnel. Resignation and resilience are both powerful mechanisms for the integration of rules and generators of subjective bias, such as the belief that equipping ourselves with these connected technologies will improve our understanding of what is happening to them and enhance the discussion held with the doctor, in order to avoid disease or, at the very least, hold it at bay. However, this involves forgetting that, in the context of the expansion of connected devices and objects, both doctors and patients become the agents and administrators of these infrastructures. The old dream of technology assisting humans functionally and occasionally in solving problems (as in the era of expert systems) is collapsing in the face of an inverted reality, in which humans are now at the service of these connected devices and objects – acting for the benefit of the device producers and the health policies they already help shape. Digital devices end up monitoring individuals from the beginning to the end of their lives, in order to relieve the burden on overloaded health services. From this, health systems emerge that aim to reduce costs, both for public health services and private insurers.

In 10 or 15 years, will connected objects and devices and AI have become essential to operating methods, just as the Internet has become essential to our social lives? The profiling of life trajectories – and therefore health trajectories – of populations is required in the generalized use of connected digital devices within a potentially freedom-reducing social environment. This is precisely what creates the imposed and inevitable character of this type of change. Technical order is imposed even as mistrust of institutions takes root, particularly in health, leading to a questioning of scientific knowledge – the "post-truth era" – that connected devices using AI are incapable of gauging. The contemporary challenge lies here.

List of Authors

Yanita ANDONOVA
LabSIC
Université Sorbonne Paris Nord
Aubervilliers
France

Dominique CARRÉ
LabSIC
Université Sorbonne Paris Nord
Aubervilliers
France

Étienne HIEN
LabSIC-ED Erasme-USPN
RHU PsyCARE-GHU
Inserm
Université Sorbonne Paris Nord
Aubervilliers
France

Christian PAPILLOUD
Martin Luther University
Halle-Wittemberg
Germany

Sarah SANDRÉ
AP-HP
IRIS
Université Paris Cité
France

Geneviève VIDAL
LabSIC
Université Sorbonne Paris Nord
Aubervilliers
France

Index

G, H

I, L

M, N

O, P

Q, R

S, T

U, V, W

Other titles from

iSTE

in

Information Systems, Web and Pervasive Computing

2026

BRÉZILLON Patrick
Context-based Modeling of Activity in Real-World Projects

2025

ABEL Marie-Hélène, MATTA Nada, KARRAY Hedi, SAAD Inès
Ethics and Digital Transition

BATTA Priya, KUMAR Abhishek, MANOJ S. Oswalt, CHINNATHAMBI Dhaya, RAVI Srivel
Sustainable Supply Chains and Carbon Footprint Reduction: The Blockchain Advantage

DJOUAD Anaïs
The Hypermodernity Factory: Working with Relational Technologies Intellectual Technologies Set – Volume 10)

KUMAR Abhishek, BATTA Priya, MANOJ S. Oswalt, CHINNATHAMBI Dhaya, RAVI Srivel
Blockchain and the Water Supply Chain: Opportunities, Challenges and Innovations

STOCKINGER Peter
The Art of Communication: Semiotic Foundations and Strategic using Applications

2024

ALBAN Daniel, EYNAUD Philippe, RICHET Jean-Loup, VITARI Claudio
Information Systems Management: Governance, Urbanization and Alignment – 2nd edition

PINET François, BATTON-HUBERT Mireille, DESJARDIN Eric
Geographical Data Imperfection 2: Use Cases

SEDKAOUI Soraya
Automated Data Analytics: Combining Human Creativity and AI Power using ChatGPT

2023

POMEROL Jean-Charles
Action in Uncertainty: Expertise, Decision and Crisis Management

REVEST Valérie, LIOTARD Isabelle
Digital Transformation and Public Policies: Current Issues

2022

ACCART Jean-Philippe
Library Transformation Strategies

BEN REBAH Hassen, BOUKTHIR Hafedh, CHÉDEBOIS Antoine
Website Design and Development with HTML5 and CSS3

BOADA Martí, LAZARO Antonio, GIRBAU David, VILLARINO Ramón
Battery-less NFC Sensors for the Internet of Things

BRÉZILLON Patrick, TURNER Roy M.
Modeling and Use of Context in Action

CHAMOUX Jean-Pierre
The Digital Era 3: Customs and Practices

KARAM Elie
General Contractor Business Model for Smart Cities: Fundamentals and Techniques

LAURENT Sébastien-Yves
Conflicts, Crimes and Regulations in Cyberspace
(Cybersecurity Set – Volume 2)

2021

EL ASSAD Safwan, BARBA Dominique
Digital Communications 1: Fundamentals and Techniques
Digital Communications 2: Directed and Practical Work

GAUDIN Thierry, MAUREL Marie-Christine, POMEROL Jean-Charles
Chance, Calculation and Life

LE DEUFF Olivier
Hyperdocumentation
(Intellectual Technologies Set – Volume 9)

PÉLISSIER Maud
Cultural Commons in the Digital Ecosystem
(Intellectual Technologies Set – Volume 8)

2020

CLIQUET Gérard, with the collaboration of BARAY Jérôme
Location-Based Marketing: Geomarketing and Geolocation

DE FRÉMINVILLE Marie
Cybersecurity and Decision Makers: Data Security and Digital Trust

GEORGE Éric
Digitalization of Society and Socio-political Issues 2: Digital, Information and Research

HELALI Saida
Systems and Network Infrastructure Integration

LOISEAU Hugo, VENTRE Daniel, ADEN Hartmut
Cybersecurity in Humanities and Social Sciences: A Research Methods Approach (Cybersecurity Set – Volume 1)

SEDKAOUI Soraya, KHELFAOUI Mounia
Sharing Economy and Big Data Analytics

SCHMITT Églantine
Big Data: An Art of Decision Making (Intellectual Technologies Set – Volume 7)

2019

ALBAN Daniel, EYNAUD Philippe, MALAURENT Julien, RICHET Jean-Loup, VITARI Claudio
Information Systems Management: Governance, Urbanization and Alignment

AUGEY Dominique, with the collaboration of ALCARAZ Marina
Digital Information Ecosystems: Smart Press

BATTON-HUBERT Mireille, DESJARDIN Eric, PINET François
Geographic Data Imperfection 1: From Theory to Applications

BRIQUET-DUHAZÉ Sophie, TURCOTTE Catherine
From Reading-Writing Research to Practice

BROCHARD Luigi, KAMATH Vinod, CORBALAN Julita, HOLLAND Scott, MITTELBACH Walter, OTT Michael
Energy-Efficient Computing and Data Centers

CHAMOUX Jean-Pierre
The Digital Era 2: Political Economy Revisited

COCHARD Gérard-Michel
Introduction to Stochastic Processes and Simulation

DUONG Véronique
SEO Management: Methods and Techniques to Achieve Success

GAUCHEREL Cédric, GOUYON Pierre-Henri, DESSALLES Jean-Louis
Information, The Hidden Side of Life

GEORGE Éric
Digitalization of Society and Socio-political Issues 1: Digital, Communication and Culture

GHLALA Riadh
Analytic SQL in SQL Server 2014/2016

JANIER Mathilde, SAINT-DIZIER Patrick
Argument Mining: Linguistic Foundations

SOURIS Marc
Epidemiology and Geography: Principles, Methods and Tools of Spatial Analysis

TOUNSI Wiem
Cyber-Vigilance and Digital Trust: Cyber Security in the Era of Cloud Computing and IoT

2018

ARDUIN Pierre-Emmanuel
Insider Threats
(Advances in Information Systems Set – Volume 10)

CARMÈS Maryse
Digital Organizations Manufacturing: Scripts, Performativity and Semiopolitics
(Intellectual Technologies Set – Volume 5)

CARRÉ Dominique, VIDAL Geneviève
Hyperconnectivity: Economical, Social and Environmental Challenges
(Computing and Connected Society Set – Volume 3)

CHAMOUX Jean-Pierre
The Digital Era 1: Big Data Stakes

DOUAY Nicolas
Urban Planning in the Digital Age
(Intellectual Technologies Set – Volume 6)

FABRE Renaud, BENSOUSSAN Alain
The Digital Factory for Knowledge: Production and Validation of Scientific Results

GAUDIN Thierry, LACROIX Dominique, MAUREL Marie-Christine, POMEROL Jean-Charles
Life Sciences, Information Sciences

GAYARD Laurent
Darknet: Geopolitics and Uses
(Computing and Connected Society Set – Volume 2)

IAFRATE Fernando
Artificial Intelligence and Big Data: The Birth of a New Intelligence
(Advances in Information Systems Set – Volume 8)

LE DEUFF Olivier
Digital Humanities: History and Development
(Intellectual Technologies Set – Volume 4)

MANDRAN Nadine
Traceable Human Experiment Design Research: Theoretical Model and Practical Guide
(Advances in Information Systems Set – Volume 9)

PIVERT Olivier
NoSQL Data Models: Trends and Challenges

ROCHET Claude
Smart Cities: Reality or Fiction

SALEH Imad, AMMI, Mehdi, SZONIECKY Samuel
Challenges of the Internet of Things: Technology, Use, Ethics
(Digital Tools and Uses Set – Volume 7)

SAUVAGNARGUES Sophie
Decision-making in Crisis Situations: Research and Innovation for Optimal Training

SEDKAOUI Soraya
Data Analytics and Big Data

SZONIECKY Samuel
Ecosystems Knowledge: Modeling and Analysis Method for Information and Communication
(Digital Tools and Uses Set – Volume 6)

2017

BOUHAÏ Nasreddine, SALEH Imad
Internet of Things: Evolutions and Innovations
(Digital Tools and Uses Set – Volume 4)

DUONG Véronique
Baidu SEO: Challenges and Intricacies of Marketing in China

LESAS Anne-Marie, MIRANDA Serge
The Art and Science of NFC Programming
(Intellectual Technologies Set – Volume 3)

LIEM André
Prospective Ergonomics
(Human-Machine Interaction Set – Volume 4)

MARSAULT Xavier
Eco-generative Design for Early Stages of Architecture
(Architecture and Computer Science Set – Volume 1)

REYES-GARCIA Everardo
The Image-Interface: Graphical Supports for Visual Information
(Digital Tools and Uses Set – Volume 3)

REYES-GARCIA Everardo, BOUHAÏ Nasreddine
Designing Interactive Hypermedia Systems
(Digital Tools and Uses Set – Volume 2)

SAÏD Karim, BAHRI KORBI Fadia
Asymmetric Alliances and Information Systems:Issues and Prospects
(Advances in Information Systems Set – Volume 7)

SZONIECKY Samuel, BOUHAÏ Nasreddine
Collective Intelligence and Digital Archives: Towards Knowledge Ecosystems
(Digital Tools and Uses Set – Volume 1)

2016

BEN CHOUIKHA Mona
Organizational Design for Knowledge Management

BERTOLO David
Interactions on Digital Tablets in the Context of 3D Geometry Learning
(Human-Machine Interaction Set – Volume 2)

BOUVARD Patricia, SUZANNE Hervé
Collective Intelligence Development in Business

EL FALLAH SEGHROUCHNI Amal, ISHIKAWA Fuyuki, HÉRAULT Laurent, TOKUDA Hideyuki
Enablers for Smart Cities

FABRE Renaud, in collaboration with MESSERSCHMIDT-MARIET Quentin, HOLVOET Margot
New Challenges for Knowledge

GAUDIELLO Ilaria, ZIBETTI Elisabetta
Learning Robotics, with Robotics, by Robotics
(Human-Machine Interaction Set – Volume 3)

HENROTIN Joseph
The Art of War in the Network Age
(Intellectual Technologies Set – Volume 1)

KITAJIMA Munéo
Memory and Action Selection in Human–Machine Interaction
(Human–Machine Interaction Set – Volume 1)

LAGRAÑA Fernando
E-mail and Behavioral Changes: Uses and Misuses of Electronic Communications

LEIGNEL Jean-Louis, UNGARO Thierry, STAAR Adrien
Digital Transformation
(Advances in Information Systems Set – Volume 6)

MONINO Jean-Louis, SEDKAOUI Soraya
Big Data, Open Data and Data Development
(Smart Innovation Set – Volume 3)

NOYER Jean-Max
Transformation of Collective Intelligences
(Intellectual Technologies Set – Volume 2)

VENTRE Daniel
Information Warfare – 2nd edition

VITALIS André
The Uncertain Digital Revolution
(Computing and Connected Society Set – Volume 1)

2015

ARDUIN Pierre-Emmanuel, GRUNDSTEIN Michel, ROSENTHAL-SABROUX Camille
Information and Knowledge System
(Advances in Information Systems Set – Volume 2)

BÉRANGER Jérôme
Medical Information Systems Ethics

BRONNER Gérald
Belief and Misbelief Asymmetry on the Internet

IAFRATE Fernando
From Big Data to Smart Data
(Advances in Information Systems Set – Volume 1)

KRICHEN Saoussen, BEN JOUIDA Sihem
Supply Chain Management and its Applications in Computer Science

NEGRE Elsa
Information and Recommender Systems
(Advances in Information Systems Set – Volume 4)

POMEROL Jean-Charles, EPELBOIN Yves, THOURY Claire
MOOCs

SAAD Inès, ROSENTHAL-SABROUX Camille, GARGOURI Faïez
Information Systems for Knowledge Management

SALLES Maryse
Decision-Making and the Information System
(Advances in Information Systems Set – Volume 3)

SAMARA Tarek
ERP and Information Systems: Integration or Disintegration
(Advances in Information Systems Set – Volume 5)

2014

DINET Jérôme
Information Retrieval in Digital Environments

DUBÉ Jean, LEGROS Diègo
Spatial Econometrics Using Microdata

HÉNO Raphaële, CHANDELIER Laure
3D Modeling of Buildings: Outstanding Sites

KEMBELLEC Gérald, CHARTRON Ghislaine, SALEH Imad
Recommender Systems

MATHIAN Hélène, SANDERS Lena
Spatio-temporal Approaches: Geographic Objects and Change Process

PLANTIN Jean-Christophe
Participatory Mapping

RIGO Michel
Formal Languages, Automata and Numeration Systems 1: Introduction to Combinatorics on Words
Formal Languages, Automata and Numeration Systems 2: Applications to Recognizability and Decidability

TORRES-MORENO Juan-Manuel
Automatic Text Summarization

VENTRE Daniel
Chinese Cybersecurity and Defense

2013

BERNIK Igor
Cybercrime and Cyberwarfare

CAPET Philippe, DELAVALLADE Thomas
Information Evaluation

LEBRATY Jean-Fabrice, LOBRE-LEBRATY Katia
Crowdsourcing: One Step Beyond

SALLABERRY Christian
Geographical Information Retrieval in Textual Corpora

TURENNE Nicolas
Knowledge Needs and Information Extraction: Towards an Artificial Consciousness

2012

BUCHER Bénédicte, LE BER Florence
Innovative Software Development in GIS

GAUSSIER Eric, YVON François
Textual Information Access

STOCKINGER Peter
Audiovisual Archives: Digital Text and Discourse Analysis

VENTRE Daniel
Cyber Conflict

2011

BANOS Arnaud, THÉVENIN Thomas
Geographical Information and Urban Transport Systems

DAUPHINÉ André
Fractal Geography

LEMBERGER Pirmin, MOREL Mederic
Managing Complexity of Information Systems

STOCKINGER Peter
Introduction to Audiovisual Archives

STOCKINGER Peter
Digital Audiovisual Archives

VENTRE Daniel
Cyberwar and Information Warfare

2010

BONNET Pierre
Enterprise Data Governance

BRUNET Roger
Sustainable Geography

BUFFET Olivier, SIGAUD Olivier
Markov Decision Processes in Artificial Intelligence

CARREGA Pierre
Geographical Information and Climatology

CAUVIN Colette, ESCOBAR Francisco, SERRADJ Aziz
Thematic Cartography – 3-volume series
Thematic Cartography and Transformations – Volume 1
Cartography and the Impact of the Quantitative Revolution – Volume 2
New Approaches in Thematic Cartography – Volume 3

LANGLOIS Patrice
Simulation of Complex Systems in GIS

MATHIS Philippe
Graphs and Networks – 2nd edition

THÉRIAULT Marius, DES ROSIERS François
Modeling Urban Dynamics

2009

BONNET Pierre, DETAVERNIER Jean-Michel, VAUQUIER Dominique
Sustainable IT Architecture: the Progressive Way of Overhauling Information Systems with SOA

PAPY Fabrice
Information Science

RIVARD François, ABOU HARB Georges, MERET Philippe
The Transverse Information System

ROCHE Stéphane, CARON Claude
Organizational Facets of GIS

VENTRE Daniel
Information Warfare

2008

BRUGNOT Gérard
Spatial Management of Risks

FINKE Gerd
Operations Research and Networks

GUERMOND Yves
Modeling Process in Geography

KANEVSKI Michael
Advanced Mapping of Environmental Data

MANOUVRIER Bernard, LAURENT Ménard
Application Integration: EAI, B2B, BPM and SOA

PAPY Fabrice
Digital Libraries

2007

DOBESCH Hartwig, DUMOLARD Pierre, DYRAS Izabela
Spatial Interpolation for Climate Data

SANDERS Lena
Models in Spatial Analysis

2006

CLIQUET Gérard
Geomarketing

CORNIOU Jean-Pierre
Looking Back and Going Forward in IT

DEVILLERS Rodolphe, JEANSOULIN Robert
Fundamentals of Spatial Data Quality

Printed and bound by CPI Group (UK) Ltd, Croydon, CR0 4YY
13/09/2026
14961048-0002